The Plantation

SOUTHERN CLASSICS SERIES

Mark M. Smith and Peggy G. Hargis, Series Editors

The Plantation ✻

Edgar Tristram Thompson

Edited with an Introduction by
Sidney W. Mintz and George Baca

The University of South Carolina Press
Published in Cooperation with the Institute for
Southern Studies of the University of South Carolina

Published by the University of South Carolina Press,
Columbia, South Carolina 29208

www.sc.edu/uscpress

Manufactured in the United States of America

19 18 17 16 15 14 13 12 11 10   10 9 8 7 6 5 4 3 2 1

Library of Congress Cataloging-in-Publication Data

Thompson, Edgar T. (Edgar Tristram), 1900–1989.
  The plantation / Edgar Tristram Thompson ; edited with an introduction
by Sidney W. Mintz and George Baca.
    p. cm. — (Southern classics series)
  Summary: First full publication of Edgar Thompson's 1932 dissertation
on the economics of the plantation.
  "Published in Cooperation with the Institute for Southern Studies of
the University of South Carolina."
  Includes bibliographical references and index.
  ISBN 978-1-57003-940-9 (cloth : alk. paper) — ISBN 978-1-57003-941-6
(pbk : alk. paper)
  1. Plantations. 2. Plantations—Economic aspects—Southern States. 3. Plantations—
Economic aspects—Virginia. 4. Land tenure—Southern States. 5. Land tenure—
Virginia. 6. Southern States—Economic conditions—19th century. 7. Virginia—
Economic conditions—19th century. I. Mintz, Sidney Wilfred, 1922– II. Baca, George.
III. University of South Carolina. Institute for Southern Studies. IV. Title. V. Series:
Southern classics series.

HD1471.A3T49 2010
307.72—dc22

2010017287

This book was printed on Glatfelter Natures, a recycled paper with 30 percent
postconsumer waste content.

Publication of the Southern Classics series is made possible
in part by the generous support of the Watson-Brown Foundation.

Contents

Series Editors' Preface   vii

Introduction   ix

1 ✴ **The Plantation as a Social Institution**   1
Introduction   1
The Plantation Defined   3
The Plantation and Colonization   4
The Plantation as a Type of Settlement   8
The Plantation and Labor   11
The Plantation as a Political Institution   13
The Theory of the Plantation   15
The Plantation and Social Change   18
Virginia as a Typical Plantation Frontier   20

2 ✴ **The Metropolis and the Plantation**   23
The Revolution in Distance   23
The Trading Factory   26
His Majesty's Plantations   33

3 ✴ **The Plantation in Virginia**   39
Free Land and Plantation Settlement   39
Agricultural Specialization: Tobacco   49

4 ✴ **Plantation Management and Imported Labor in Virginia**   56
The Tide of White Labor   56
Negro Slavery and Its Control   63
The Evolution of the Planter   71
The Humanization of the Plantation   74

5 ✴ **The Plantation and the Frontier**   82
Economic Changes and the Small Farm in Virginia   82
The Plantation on the New Southern Frontier   87

6 ✒ **The Natural History of the Plantation   100**
  Geographical Isolation and Culture   100
  Ecological Changes and Race Relations   103
  Adaptation and Accommodation to a New Habitat   104
  Agricultural Specialization and Racial Stratification   107
  The Organization and Control of Labor   108
  Peasant Proprietorship and Cultural Homogeneity   110

  Notes   113
  Bibliography   137
  Index   147

Series Editors' Preface

Edgar T. Thompson, a southerner by birth and a sociologist by training, recast childhood experiences on his father's plantation to fuel an intellectual journey that placed the plantation at the analytical center of his sociological investigations. In this, the publication of Thompson's doctoral thesis in its entirety, we come to understand plantation agriculture, southern exceptionalism, and black/white race relations as parts of the larger enterprises of European state building and global capitalism. Students of the *new* global South will benefit from this early attempt to link locality to global networks as they are reminded that the South's ties to a global political economy predated the Civil War.

Southern Classics returns to general circulation books of importance dealing with the history and culture of the American South. Sponsored by the Institute for Southern Studies, the series is advised by a board of distinguished scholars who suggest titles and editors of individual volumes to the series editors and help establish priorities in publication.

Chronological age alone does not determine a title's designation as a Southern Classic. The criteria also include significance in contributing to a broad understanding of the region, timeliness in relation to events and moments of peculiar interest to the American South, usefulness in the classroom, and suitability for inclusion in personal and institutional collections on the region.

MARK M. SMITH
PEGGY G. HARGIS
Series Editors

Introduction

The publication in its entirety of *The Plantation,* Edgar T. Thompson's doctoral thesis, is particularly timely. Completed seventy-eight years ago, it constitutes a pioneering approach to the study of early capitalistic experiments in overseas export-oriented tropical agriculture that used forced labor on land taken from native peoples, with capital, plants, and technology coming from Europe and Asia. Except for its first chapter, it has never been published. As an important document in American intellectual history, as well as in the history of the so-called Chicago School of sociology, it stands on its own.

We call its publication timely because of recent radical changes in the shape of the world. The last decades of the twentieth century and the first of the twenty-first were marked by a widely shared consciousness of the growing importance of globalization. The sustained volume and, soon enough, velocity of movement—of people, of commodities, and of capital—had given rise to dubiously optimistic expectations about what might happen next. Explosive new forms of communication, barely imagined before, were beginning to overwhelm the leadership of even large and repressive states. In economic terms business decisions were being made, and then acted upon, with what seemed to be runaway speed. The growing assumption that such changes in the world were symbols of a wholesale globalization was abetted by the apparent unawareness that there had been other globalizations, not so many years earlier, that had taken shape and then broken apart.[1] Various attempts by both anthropologists and historians to compensate for widespread failings of historical memory by counseling a broader, more open world-historical approach received relatively little attention.[2]

As one looks back now, a longer, less occluded historical outlook seems called for. Obviously the chances for any consensus on the fate of globalization and the significance of recent history will remain slight for at least another half century, but the realization that depressions, as well as globalizations, were actually familiar phenomena long before now—indeed that they were phenomena lived through by a great many people still alive—is helping to bring attention to this past. Long before now, some people realized that the current globalization had predecessors: they recognized, for instance, that the American South had become part of a wider world before the Civil War. Now there is a renewed inclination to look back while confronting head-on the idea that other globalities preceded this one.

Thompson's doctoral thesis, along with the articles and books that were to follow during a long scholarly life, represents one of the earliest attempts to reinterpret the history of the American South as an integral part of global processes. In what he referred to as stages toward the creation of a "world community," Thompson showed how southern exceptionalism and the regional obsession with race, which took shape as early as the seventeenth century, were actually intertwined in the rise of the European state. The coalescent industrial and economic systems those states represented were not divorced from their expanding colonial policies. Thompson's grasp of this wider ensemble of forces led him to reconceptualize the plantation as a *political* institution. By means of the large-scale, quasi-industrial production of agricultural staples abroad—staples that served to absorb the rising buying power of consumers in the home metropolis—the plantations contributed to the international power of European states. By seeing this inside politico-economic connection, Thompson was also able to see that the plantations, lying outside Europe but lodged in European colonies and ex-colonies, were, as he wrote, "race making" institutions as well.[3] Put simply, plantations not only produced what were once costly foods on the cheap; their labor systems also sorted colonized and colonial peoples socially. Focusing upon the ties that bound conquest and state power to the reification of racial categories, Thompson was able to show how the plantation's existence had helped to articulate Europe, Africa, and the New World politically, economically, and largely on the colonial masters' terms. As pioneering institutions in frontier areas, plantations represented the deepest penetration of European power. Once locally installed, however, the plantation regime could become antithetical to the state, even while ready to enlist its support in conniving to maintain local control.[4]

Intellectual Formation

Edgar Thompson's research into the global significance of plantations clearly drew upon childhood experiences. Born in 1900, he had grown up on his father's small (and moribund) plantation in Dillon, South Carolina, just south of the North Carolina border.[5] It was years, however, before he grasped fully the relationship of the daily routines of plantation life he knew personally to the larger historical questions that motivated him intellectually. At the time he left the plantation to attend the University of South Carolina in Columbia, he thought the plantation institution had little historical weight.[6] Like all of us who have trouble objectifying a lived childhood experience, Thompson was inclined to take plantation life and its attached cultural values for granted.

After graduating from college, he accepted a job teaching rural sociology at the University of North Carolina at Chapel Hill. After a year there, he was offered a permanent post in the department on the condition that he earn a master's degree at a university in either the Midwest or the North. He completed that degree at the University of Missouri,[7] but then he became frustrated after returning to Chapel Hill and departed to become a doctoral candidate in sociology at the University of Chicago. In the late 1920s Chicago's sociology department was a thriving center for research on the modern world, and there Thompson became the student of Robert E. Park, the sociological pioneer. Park inspired Thompson intellectually, eventually convincing him that the plantation, when viewed through time and in its different guises, was a suitable subject for his doctoral dissertation. It was at Chicago that Thompson began his lifelong reflection upon the plantation as a global phenomenon.

Park, his teacher, had worked as a newspaper reporter in the South and, curiously, as Booker T. Washington's ghostwriter. These experiences had left him with a keen interest in the topics of racism and the American South. Over the years, Park attracted and cultivated a group of highly promising students. Among them were sharp critics of southern agriculture and race relations, including E. Franklin Frazier and Charles S. Johnson, who became prominent sociologists. Park also worked with William Oscar Brown, another white southerner, who wrote on race prejudice in Texas, and with Everett C. and Helen MacGill Hughes.

Park's work convinced him that a new period in American history was taking shape, and he urged his students to analyze "the historical process by which civilization, not merely here but elsewhere, has evolved, drawing into the circle of influence an ever widening circle of races and people."[8] Park became famous in the social sciences for encouraging his students (among them the young anthropology student Robert Redfield) to break down for scrutiny the phenomenon of modernity, into which their research was drawing them.

That two outstanding African American scholars were among his students so early in the fight for civil equality meant that Park was able to play a pioneering role in transforming the composition of the American academy. Like his African American colleagues, Thompson used Park's ideas to study the forces that intertwined with plantation agriculture, race, and the history of slavery in defining the South. As he did so, he was attracted to the possibilities of objective inquiry into the region's seemingly intractable problems, including racism and the agrarian economy. In today's academic world, Thompson's belief in the ideals of scientific objectivity may seem quaint and naive. Viewed instead from the perspective of the Jim Crow South as Thompson knew it firsthand, the scientific study of the plantation would enable him to examine more objectively the structure of that society. Thompson was able to define his scholarship carefully to avoid directly challenging white supremacy or the deep local anxieties about racial equality. Yet by using the plantation as his lens, he was able to ask some of the previously unspoken questions about hierarchy and social behavior. Those who never lived in the Jim Crow South may have difficulty conceding as much, but Thompson's historical studies of the plantation led him to stances on racial integration and interracial social relations that were remarkably progressive eight decades ago.

Thompson used his conception of the plantation to situate the life of the South within far larger, global processes: colonization and colonialism in the New World, the slave trade and slavery, and the maturation of global capitalism. In anthropology Robert Redfield would use some of the ideas of the Chicago group to carry out a much more elegant (but ultimately, perhaps, less fruitful) analysis.[9] In contrast, by depicting plantations as devices that contributed to far-reaching changes of a kind never envisaged by the societies in which they flowered, Thompson aligned himself (likely to some extent unknowingly) with other social critics of modernity, including Karl Marx, Max Weber, and Werner Sombart.

After receiving his degree from the University of Chicago in 1932, Thompson began teaching at Skidmore College. The following year he joined the sociology department at Duke University. He continued to teach there until 1970 and founded Duke's Center for Southern Studies. He was also instrumental in developing the Black Studies Program there.

The Plantation

Thompson's doctoral thesis is a document in intellectual history. It is also a guidepost of a sort. For those interested in how the modern world grew, particularly in relation to agriculture, or in what ways the effects of modernization can be uncovered in the character of the contemporary South, Thompson's work has much to say. To be sure, his materials are in many ways dated or incomplete. But his grasp, more than seventy-five years ago, of how Europe was tied to the American South by the important role that cotton plantations played in European industry may stir in today's students a sense of the South's place in the global economy. Reading his chapters on the internal development of European society—a development that resulted, inter alia, in the rise of modern plantations—one grasps that Thompson did not see the South as self-contained or as enchanted by some nostalgic vision of its own past.

Thompson himself successfully avoided the American exceptionalism championed by some northern scholars and the regional parochialism of many southern historians by reflecting on the research of early anthropologists such as Henry Maine, Lewis Henry Morgan, and, at later points, Franz Boas. The relevance of their work became clearer as the social sciences continued to change. At the very time that Thompson was writing his thesis, the distinction between anthropology as the study of "the primitive" and sociology as the study of "modern life" took firmer shape. Thompson managed to ignore that arbitrary and "gentrified" distinction in producing what he conceived of as a "natural history" of the plantation. Implicit in his analysis is the recognition that Western institutions such as the plantation had emerged in the murky spaces that seemed to lie between "the primitive" and "the modern" and that were consigned to an unexamined limbo by the barriers then being built between anthropology and sociology.

Thompson also came to understand how the plantation had battened upon slavery, a supposedly "archaic" institution, reviving and reinstituting it in the West, particularly in the tropical Americas. He came around to interpret modern forms of racism as a product of plantation labor history. He believed in the objective existence of racial groups, but his scholarship focused on the way the plantation system imputed racial characteristics and exaggerated racial differences as part of its struggle to control and secure a dependable labor force. In this way Thompson described how European planters, living in colonial communities dedicated to the overseas production of agricultural staples, made use of those hierarchical conceptions of socioracial classification that plantation life had nourished and helped to spread. He made this connection between racial ideas and the functioning of power on the plantation at the time when the anthropology of Franz Boas promised a genuinely scientific approach to race.[10] Thompson's careful separation of biological from cultural criteria of difference threw light upon the ways in which culturally invasive European institutions stigmatized non-Europeans, particularly in the tropical regions of the Americas.

New ideas about colonial governance would lie fallow for several decades while anthropology continued to refine itself as the discipline that studied "primitives."[11] But after World War II, North American anthropology changed rapidly. Returning from their experiences in the Spanish Civil War, World War II, and Korea, a new generation of students were entering graduate school and beginning to test the anthropological canons of their teachers. Some were looking for approaches that would let them study problems that went beyond the existing boundaries of the discipline. Subjects such as modern commercial agriculture, ethnicity, and peasantries were for the most part not suitable topics of study before World War II.[12] It was in the postwar period of decolonization and postcolonial nation building that it would become fashionable to study slavery, colonization, colonial empires, and even modern nation-states.

One of the better-known attempts to expand the anthropological approach to the modern world was Julian Steward's People of Puerto Rico Project. In attempting to apply anthropological methods to the study of a large Western society,

Steward recruited graduate students from Columbia, the University of Puerto Rico, and the University of Chicago. He asked each to select specific local communities to study in Puerto Rico that would represent "major economic adaptations." His aim was to develop a conceptual framework that would make possible the comparison of communities, so as to lay bare—as he saw it—the interwoven institutions that knitted those communities together into a nation.[13]

The resulting book, *The People of Puerto Rico* (1956), failed to solve the difficult methodological and theoretical problems posed by national institutions.[14] As a cooperative undertaking that enabled simultaneous study of economically different communities within a single society, however, the volume marked a methodological turning point. Inevitably the disciplinary expansion of anthropology's scope moved the meaning of "community" away from the older sense of a society that could be explained in terms of itself and toward its redefinition as a "working part" within larger economic and political networks.

But this also raised questions for which the discipline was unprepared. Anthropologists working in plantation and peasant communities in the Caribbean, Africa, Latin America, and the Indian Ocean, for example, found themselves writing colonial history and learning the histories of agricultural staples such as coffee, sugar, and tobacco. Students who had been trained to study communities as if they were tiny islands or isolated tribes found themselves discovering instead how tropical regions had earlier been thrust into an ahistorical "non-Western" category. Inevitably some began to rethink how the postwar processes of decolonization and nation building had taken shape. They became aware of the reluctance of landed colonial elites to surrender power and of the ambivalence of empire.

For many there was a sense that anthropology had repudiated its role as a discipline meant to study "primitive people." But for those who now wanted to know what had become of the peoples once called "primitive," there seemed to be no turning back. Thompson's work is relevant to this connecting of ethnographic research to a more inclusive world history.

Thompson himself was not skilled at self-promotion. In 1945, more than a decade after his thesis was accepted, only one chapter of it had ever been published. He was by then a tenured professor—and still learning. He wanted to see more comprehensive plantation research, but he had not attracted a student following. He knew that his deep interest in plantations as a global phenomenon was viewed by many of his contemporaries as merely antiquarian. In the decades before World War II, it would have been highly unconventional for any social scientist to undertake research on the United Fruit Company estates in Central America or any of the big pineapple or sugar plantations around the world. A serious anthropological study of henequen plantations in Yucatán, for instance, would not come for another decade or so; and Redfield's many books on the peninsula had ignored them.[15] For his part Thompson seemed quite content to remain outside the professional mainstream of social science theory, apparently wanting most of all to tend to his own theoretical garden. Today, however, those who read him carefully will discover that garden was well tended.

Frontier Institutions: From Inside Out

Though Thompson knew the plantation milieu of his childhood, he chose to begin his exploration of agrarian labor history in places remote in both time and space: ancient Greece, East Prussia, and the factory system of sixteenth-century England. His thesis was heavily theoretical. Throughout he firmly attached his research on the southern plantation to the theory of the state and the rise of colonial systems. He used the plantation in sketching specific links between Britain, on the one hand, and Africa, the American South, and the Caribbean, on the other. He related the plantation to the theory of the state by reading Herman Jeremias Nieboer, Lewis Henry Morgan, and Franz Oppenheimer.[16] Drawing on theories of state power, Thompson neatly detached the southern plantation from its naturalizing myths—and from the protective provincialism of wistful southerners—to place it at the very center of expanding North Atlantic capitalism. Indeed he conceived of the planter as an authoritarian figure, antithetical to the modernity represented by the European state, since firmly established plantation regimes typically blunted the spread of civil society, state institutions, and abstract governance. But Thompson drew our attention to a less obvious effect of the plantation: its guise as a political institution that served the process of state making, by "bringing men under new and more stringent forms of control."[17] His "theory of the plantation" was meant to clarify the process of class formation in societies typified by powerful forced-labor institutions.

He employed the plantation to theorize specific global connections that transformed the globe into what he called "a world community." Subsequent thinkers, building on his work, renamed that world community North Atlantic capitalism.[18] Thompson grasped the significance of the plantation's prominent role as a "frontier institution." In its beginning phase, as a frontier institution, the plantation is put to work by colonists and mercantilists to forge "new and higher uses" for the natural resources of just-discovered or thinly settled regions. Focusing mostly on England and its rise as an imperial maritime power, he showed how the plantation had been historically important to the consolidation of the modern state. Because of its productive power, it was tied to the industrialization of Europe, yet largely compliant with the early imperial insistence upon total economic subordination of the colonies.[19] Historians and social scientists who are drawn to the alleged newness of globalization will benefit from reading this early attempt to make concrete and specific the links of locality to global networks and the consequences of those ties.

Power and Ideas

In addition to its relevance to debates about globalization, *The Plantation* offers insights into how ideas can serve to sustain institutional order. Since the late 1970s American anthropology has entertained the notion that social categories are fluid and in a state of constant flux, such that their meanings, accordingly, are similarly motile. Many contemporary writers now view this idea as a distinctive aspect of our postmodern world. Thompson's institutional approach to the plantation suggests that fluidity and change are in fact not new at all. He envisioned the plantation as representing the larger struggle of European states to enforce order—at home as well as in the colonies—even as it creates disorder and uncertainty by tearing asunder group relations and leaving indigenous or enslaved people disorganized and unattached. Indeed it was by building upon the systematic destruction of preexisting social forms and kin groups that modern states sought to produce and enforce order, and plantations were often the farthest outposts of the state. Order, then, becomes intricately entwined with institution building by the modern state and its expanding frontiers. Drawing insight from the description of the Greek polis and from the ideas of Frederick Jackson Turner, the historian of the American frontier, Thompson argued that the early settlement period that gives rise to the plantation is, by analogy, one of "turmoil and disorder," and that the plantation, much like the Greek polis or the European state, "arises as a means of establishing order, of restoring 'the intricate web of normal expectations,' and, unlike the *Polis,* of maintaining industrial and market relations. This double function of the plantation leads to enslavement or other forms of forced labor, to the importation of more individualized, or familyless labor, and to concentrated rather than dispersed settlement."[20]

Thompson's work has genuine implications for contemporary debates about racial politics. After the successes of the U.S. civil rights movement and of anticolonial struggles throughout the world, many overtly racist policies were dismantled. Yet after four decades of racial reform, racial institutions have changed, leading many scholars to see contemporary forms of racism as "contradictory," "fluid," and "complex." For Thompson such mutability of racial ideas stem from their institutionalization through the plantation.[21] Foreshadowing Barbara Fields's insightful work by decades,[22] Thompson wrote that the plantation was "a race-making situation."[23] He would have shared Fields's dismay at the way in which modern scholars wield the fashionable slogan of race as "socially constructed" without specifying the political uses of that perspective. *The Plantation* deals with the specific uses of racial ideas for the expansion of the frontier and the production of agricultural staples for global markets. "In an area of open resources" (such as the frontier in the West Indies and Virginia), Thompson pointed out, "it is necessary to control and incidentally coerce labor in order to bring land into new and 'higher' uses. Slave labor is likewise a form of capital with which surplus is produced to bring more land into new and higher uses."[24] African servants gradually displaced English servants. With the emergence of tobacco as an agricultural staple in Virginia, he notes that the "competitive opportunities of the Negro were restricted with greater ease and acceleration than in the case of the white servants." As a result a conception developed by "each group of the other as members of a separate and distinct 'race,' a different species, each biologically equipped with immutable instincts and dispositions constituting the one into 'higher' and the other into a 'lower' race."[25] In the end "race" was an economic relation that underpinned the New World plantation. He conceived of race as a relationship that integrated people of African and European descent into a single, but hierarchically divided, community. He focused on what blacks and whites shared, illustrating that African Americans did not hold contrary values.

Conclusion

We asserted at the outset that Thompson's work is both relevant to contemporary research and a document in the intellectual history of the social sciences. Some modern scholars, including historians such as Sir Eric Eustace Williams and Barry Higman and anthropologists such as Eric Wolf and Michel-Rolph Trouillot, have drawn inspiration from his work. We have sought to highlight here some of the research themes that Thompson himself brought to light. We hope the reader will be moved to look for more in the pages that follow.

Editorial Method

Edgar Thompson's words are presented with as little interference on our part as possible. What we deemed obvious typos or misspellings were silently corrected, and the capitalization of terms and titles was standardized. Occasional grammatical irregularities were straightened out. We resisted the impulse to tinker with punctuation unless we felt it was needed to prevent misunderstanding. In the notes, however, we substituted short citations whenever possible and provided full source citations in the bibliography.

Acknowledgments

We wish to express our appreciation to colleagues who read and criticized earlier drafts of this introduction. Professors David Nugent, John Shelton Reed, and Mark Smith read our initial effort and provided helpful advice and encouragement. Professors Alex Lichtenstein and Marc Edelman made thorough reviews of a later draft and provided valuable guidance with their criticisms. We warmly thank them all. Finally we thank Beverly Phillips of the Steenbock Memorial Library at the University of Wisconsin–Madison. She provided help at a crucial moment in the most generous fashion.

Notes

1. Frederick Cooper, "What Is the Concept of Globalization Good For? An African Historian's Perspective," *African Affairs* 100 (April 2001): 189–213.

2. Advocates of such an approach include Thomas Bender, *A Nation among Nations: America's Place in World History* (New York: Hill and Wang, 2006); Don Harrison Doyle and Marco Antonio Villela Pamplona, *Nationalism in the New World* (Athens: University of Georgia Press, 2006); and James L. Peacock, *Grounded Globalism: How the U.S. South Embraces the World* (Athens: University of Georgia Press, 2007).

3. Edgar T. Thompson, "The Plantation as a Race-Making Situation," in *Sociology: A Text with Adapted Readings,* ed. L. Broom and P. Selznick, 506–7 (New York: Harper and Row, 1955). See also Michel-Rolph Trouillot, "Culture on the Edges: Caribbean Creolization in Historical Context," in *From the Margins: Historical Anthropology and Its Futures,* ed. K. Axel, 189–210 (Durham, N.C.: Duke University Press, 2002), esp. 200.

4. Alex Lichtenstein, "Ned Cobb's Children: A New Look at White Supremacy in the Rural U.S.," *Journal of Peasant Studies* 33, no. 1 (2006): 124–39.

5. Today Dillon is most famous for its kitschy roadside attraction, the Mexican-themed South of the Border.

6. Sociology at Duke University, Box #3, Folder #56, Edgar Tristram Thompson papers, University Archives, Duke University.

7. Sociology at Duke University, Box #3, Folder #56, Edgar Tristram Thompson papers, University Archives, Duke University.

8. Robert E. Park, "An Autobiographical Note," in *Race and Culture* (Glencoe, Ill.: Free Press, 1950), vii–viii.

9. Robert Redfield, *Tepoztlan, a Mexican Village: A Study in Folk Life* (Chicago: University of Chicago Press, 1930), and Redfield, *The Folk Culture of Yucatan* (Chicago: University of Chicago Press, 1941). See also Oscar Lewis, *Life in a Mexican Village: Tepoztlán Restudied* (Urbana: University of Illinois, 1951); Sidney W. Mintz, "The Folk-Urban Continuum and the Rural Proletarian Community," *American Journal of Sociology* 59, no. 2 (1953): 136–43; Gideon Sjoberg, "Folk and Feudal Societies," *American Journal of Sociology* 58, no. 3 (1952): 231–39; and Joan Vincent, *Anthropology and Politics: Visions, Traditions, and Trends* (Tucson: University of Arizona Press, 1990), 284–92.

10. Franz Boas, "Human Faculty as Determined by Race," *Proceedings of the American Association for the Advancement of Science* 5, no. 43 (1894): 301–27.

11. For a notable exception, see two works by Hortense Powdermaker, *After Freedom: A Cultural Study in the Deep South* (New York: Viking Press, 1939), and *Stranger and Friend: The Way of an Anthropologist* (New York: W. W. Norton, 1966).

12. For pioneering work in the anthropology of commercial agriculture, see Walter Goldschmidt, *As You Sow* (New York: Harcourt, Brace, and Company, 1947), and Julian Steward et al., *The People of Puerto Rico* (Urbana: University of Illinois Press, 1956); for works in ethnicity, see Lloyd Warner, *The Social Systems of American Ethnic Groups* (New Haven, Conn.: Yale University Press, 1945); and for a work that challenged established assumptions about peasantries, see Eric R. Wolf, *Peasants,* Foundations of Modern Anthropology Series (Englewood Cliffs, N.J.: Prentice-Hall, 1966).

13. Sidney W. Mintz, "People of Puerto Rico Half a Century Later: One Author's Recollections," *Journal of Latin American Anthropology* 6, no. 2 (2002): 79.

14. For assessments of the impact of the work of Steward et al., see Sidney W. Mintz, "The Role of Puerto Rico in Modern Social Science," *Revista/Review Interamericana* 8, no. 1 (1978): 5–16; William Roseberry, "Historical Materialism and *The People of Puerto Rico,*" *Revista/Review Interamericana* 8, no. 1 (1978): 26–36; and Eric R. Wolf, "Remarks on The People of Puerto Rico," *Revista/Review Interamericana* 8, no. 1 (1978): 17–25.

15. Mintz, "The Folk-Urban Continuum."

16. Herman Jeremias Nieboer, *Slavery as an Industrial System; Ethnological Research* (New York: B. Franklin, 1901); Lewis Henry Morgan, *Ancient Society; or, Researches in the Lines of Human Progress from Savagery, through Barbarism to Civilization* (London: Macmillan, 1877); and Franz Oppenheimer, *The State: Its History and Development Viewed Sociologically* (1909; repr., New York: Vanguard Press, 1926).

17. Edgar Tristram Thompson, *The Plantation,* ed. Sidney W. Mintz and George Baca (Columbia: University of South Carolina Press, 2010), 1.

18. See Sidney W. Mintz, *Caribbean Transformations* (Baltimore: Johns Hopkins Press, 1984); Michel-Rolph Trouillot, *Global Transformations: Anthropology and the Modern World* (New York: Palgrave Macmillan, 2003); Eric R. Wolf, *Europe and the People without History* (Berkeley: University of California Press, 1981).

19. Mintz continues to employ Thompson's insights in his own work, including a volume on the Caribbean, *Three Ancient Colonies: Caribbean Themes and Variations* (Cambridge, Mass.: Harvard University Press, 2010). For a similar approach to understanding European feudalism, see Witold Kula, *An Economic Theory of Feudalism: Towards a Model of the Polish Economy, 1500–1800,* trans. Lawrence Garner (London: New Left Books, 1976).

20. Thompson, *The Plantation,* 20.

21. See Ann Stoler, "Racial Histories and Their Regimes of Truth," *Political Power and Social Theory* 11 (1997): 183–206.

22. Barbara J. Fields,"Ideology and Race in American History," in *Region, Race, and Reconstruction: Essays in Honor of C. Vann Woodward,* ed. J. M. Kousser and J. M. McPherson, 143–78 (New York: Oxford University Press, 1982), and Fields, "Slavery, Race and Ideology in the United States of America," *New Left Review* 181 (May–June 1990): 95–118.

23. Thompson, "The Plantation as a Race-Making Situation."

24. Thompson, *The Plantation,* 13.

25. Ibid., 64.

The Plantation

1

The Plantation as a Social Institution

Introduction

The problem of the plantation, whose lusty revival in tropical countries we are now witnessing, is a part of the larger problem of what Teggart calls "politicization," the process of state-making, of bringing men under new and more stringent forms of control.[1] Concerning the state there are a large number of interpretations, and many explanations of classes, castes, and forms of forced labor, such as slavery, have been advanced. But there is very little in the literature of social science that might be called a theory of the plantation. Any effort to supply such a theory will face at the outset certain fundamental questions: What is a plantation? Why does the plantation arise in some areas and not in others? What is the natural history, i.e., the cycle of change, of this institution? Concerning questions of this sort, answers have come mainly from students of colonization. These students have, in general, sought to explain the plantation mainly in terms of climatic influences or causes.[2]

It is the thesis of this study that the plantation is to be regarded as a political institution which has a natural history very much like that of other types of political institutions as, for instance, the state. The plantation, so far as it may be regarded as a political institution, is one that exists for the purpose of bringing land into new and higher uses through the medium of an agricultural staple which is sold on the metropolitan market. The plantation arises out of settlement, out of the contact and collision of diverse racial and cultural groups on a frontier, as a means of maintaining and realizing original economic purposes. It acquires its institutional characteristics in the process of meeting and finding some sort of solutions to its most persistent problems: the problem of operating at a profit and of getting and controlling an adequate supply of cheap labor. Its purposes are industrial; its means for achieving these purposes are political. It is a political institution in so far as it introduces, or evolves, and enforces order where there has been disorder and uncertainty among individuals who have been torn out of former group relations and left disorganized and unattached. The plantation arises as the personal "possession" of the planter who is able to acquire it, to enforce his authority, and incidentally, to compose conflicts and settle individuals on the land. In the process an aristocracy and a peasantry are established with appropriate attitudes of loyalty, responsibility, and control.

Amid a multitude of details and a variety of entangled considerations of general interest the theme we seek to keep clearly before the reader is that the plantation is an incident in the conquest, settlement, and exploitation of a frontier area, and that its changes mark the changes in the frontier itself. The general process in which the plantation originates and develops is designated by Teggart as politicization.

Politicization, on its objective side, seems to denote a relative transition from a form of society in which the collective force is diffuse and whose integrating principle is consanguineous or totemic, to another in which power is usurped by, or delegated to, particular individuals, or a class of individuals, and whose unity is based upon locality. For the understanding of how human institutions have come to be as they are the importance of this transition can hardly be exaggerated. Or such was the conviction of Sir Henry Maine:

> The history of political ideas begins with the assumption that kinship is the sole possible ground of community in political functions; nor is there any of those subversions of feeling, which we term emphatically revolutions, so startling and so complete as the change which is accomplished when some other principle—such as that, for instance, of *local contiguity*—establishes itself for the first time as the basis of common political action.[3]

For Teggart, the problem of politicization arose in connection with his inquiry into the proper uses of history in the study of how man and his institutions have come to be as they are, or, more abstractly, of historical events in relation to social change. It is observed that "political organization is an exceptional thing, characteristic only of certain groups, and that all peoples whatsoever have once been or still are organized on a different basis."[4] Various theories that have been proposed in explanation of the fact that political institutions are unequally distributed among the peoples of the earth are examined and rejected by Teggart. Two of these are the familiar theories of geographical and of racial determinism.[5] Teggart's point of view, as an historian, is to accept man "as given," to leave aside all questions of innate differences, and to regard change as ensuing "upon a condition of relative fixity through the interposition of shock or disturbance induced by some exterior incident."[6] In migration is sought the major source of these shocks and disturbances.

Political institutions arise at the termini of routes of travel, at "points of pressure," where collision and conflict have broken up kindred groups.[7] Collision and conflict, in breaking up the older organization, have the effect of liberating the individual man from the customary dictation of his group "as a result of the breakdown of customary modes of action and of thought, the individual experiences a 'release' in aggressive self-assertion. The overexpression of individuality is one of the marked features of all epochs of change."[8] This aggressive self-assertion and individual release

> has been the necessary prelude to the emergence of territorial organization and the institution of personal ownership. However far apart these elements may appear in modern life, in the beginning they are identical, for the fundamental characteristic of political organization is the attitude of personal ownership assumed by the ruler toward the land and the population over which he has gained control—an attitude expressed to this day in the phrases "my army" and "my people." . . .
>
> The crucial point to be observed here is that kingship and territorial organization represent simply the institutionalization of a situation which arose out of the opportunity for personal self-assertion created by the breakup of primitive organizations and it should be understood that just as the relative stability of the older units follows from the fact that every human being is born into a given group and becomes assimilated to this in speech, manners, and ideas, so, in this new organization, the *status quo* operates to perpetuate itself, and the mere fact of its existence becomes an argument for regarding it as ordained by some super-mundane power.[9]

The Plantation Defined

The plantation, as here considered, is a large landed estate, located in an area of open resources, in which social relations between diverse racial or cultural groups are based upon authority, involving the subordination of resident laborers to a planter for the purpose of producing an agricultural staple which is sold in a world market.[10] In discussing the plantation we are, of course, not limiting ourselves to any one historic institution; rather we are dealing with the generic, the typical plantation. Each historic plantation area may show unique characteristics, but the above mentioned features they all hold in common. There are a number of social and economic organizations which, while closely related to the plantation, differ in one or more important respects from it. We may notice a few of these.

The manor represents a situation where economic self-sufficiency has entered the mores as something of an ideal and rent rather than profits is sought.

Contrarily, the plantation depends upon the development of trade and transportation. But the manor actually has much the same history as the plantation. Both the lord and the planter exercise judicial functions and become eventually officials of the state.

The ranch is related to the plantation, but, because it produces a commodity which ordinarily can transport itself to market, it is usually located further inland than is possible for commercial agriculture without cheap transportation. The ranch also is an institution for bringing land into new and higher uses. The labor economy of the ranch, however, diverges sharply from that of the plantation at the point where it becomes possible by means of mounted laborers to manage large numbers of livestock. The cowboy, mounted, wild and free, has little in common with the routine stoop-laborers of the plantation.

The plantation differs from the so-called large power farm, such as is devoted to wheat cultivation on the Great Plains, by the presence of operations demanding a uniform type of unskilled labor. On the plantation, machine methods for such operation either do not exist, or are uneconomic. In the absence of adequate seasonal labor it becomes necessary for the plantation to maintain sufficient labor throughout the year to meet the requirements of the peak.

The large agricultural mission is another way of bringing land into higher uses and exhibits another aspect of the process of politicization. Like the plantation it may produce staples for sale on the market, but the agricultural mission is fundamentally an institution maintained specifically for the assimilation and education of the native. Although the plantation is frequently justified as training in "regular work and healthy exercise," we may safely say that training and education are rarely serious motives of the planter in operating a plantation. The test of plantation success or failure is the return of the investment which it gives, and the predominance of motives other than those of profit-making is sufficient to change it into some other sort of institution. The plantation governs its membership, therefore, not for the glory of God, but for the material advancement of His planters. Nevertheless, although no part of the planter's motive, the plantation is a powerful agency of assimilation and acculturation.

The Plantation and Colonization

The history of the plantation is bound up with the discovery of new lands and the expansion of commerce, with the steamship, the railway, and other new means of transportation. It is bound up with the growth of colonies and cities, and of a world market. It is, in short, a colonial institution producing for the world market what the Germans call *kolonialwaaren,* i.e., sugar, spices, etc.[11] Colonization is one way of extending the community's frontier.

Students of urban institutions and areas, the slum, for instance, have found it necessary to understand each area and institution in the wider context of the entire urban community, the area over which there is competitive cooperation. Likewise the plantation, as a colonial institution, cannot adequately be understood apart from the economic and geographical coordinates which constitute the world community.

In the world-wide economic "web of life" peoples and regions maintain more or less specialized divisions of labor in an organization which is sometimes called the Great Economy. The Great Economy creates the occupational "places," or jobs, which are relatively stable and which a succession of individuals may fill. This is one ordinate of the world community. The other is the still more stable geographical base upon which the Great Economy is overlaid. By the world community we mean that competitive and cooperative organization which is observed in the distribution of individuals, or groups of individuals, on the map, and which we seek to analyze into a system of spacial relationships between such individuals or groups of individuals. In the world community the location of institutions is seen not merely as a fact but as a problem, the problem of understanding the processes that determine their spacial position.

The accompanying map is intended to give some conception of the present distribution of plantations in the world community.[12] Plantation areas are predominantly in tropical or semi-tropical regions forming a belt about the equator all around the world. The most absolute extent to which they are fixed on continental coasts and archipelagoes in this belt indicates their relation to cheap ocean transportation. It is only by way of the ocean that any commodity enters the world market; the world market exists, for one reason, because ocean transportation is relatively so cheap that competition becomes world-wide. It is no accident, therefore, that the islands of the East and West Indies, with easy access to cheap ocean transportation, have been among the most important centers of plantation agriculture.

The tropical and semi-tropical distribution of present plantations has suggested to certain students that the explanation of the plantation may be found in climatic influences or causes. As previously stated, such is the usual explanation of students of colonization. The work of Keller may briefly be considered as representative of this interpretation.[13] Many, if not most, of the varying geographical factors influencing the form and development of colonies are, according to Keller, finally correlated with differences in climate, and climate, for all practical purposes in the study of colonization, may be broadly divided into tropical and temperate. "[A]griculture is the only important primary form of the industrial organization common to colonies of all latitudes and altitudes" and, therefore, is "the only

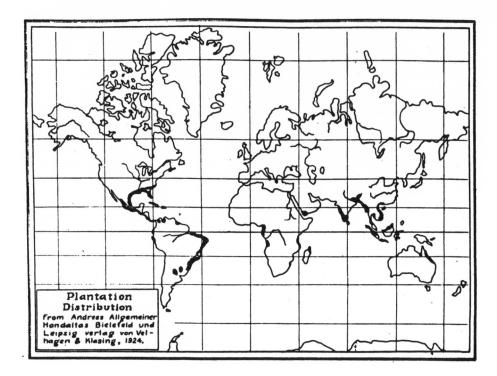

Plantation
Distribution
from Andrees Allgemeiner
Handaltas Bielefeld und
Leipzig verlag von Vel-
hagen & Klasing, 1924.

criterion of classification of adequate generality, not to mention importance."[14] Agriculture is adjusted to climatic conditions with all variations between temperate and tropical climates. Colonies based upon temperate agriculture become farm colonies, whereas those based upon tropical agriculture are characteristically plantation colonies.

The temperate farm colony is marked by economic and administrative independence, democracy, and equality. Its unit of organization is the family, and the population is fairly well divided between the sexes. Hence there is little contact with native women and no mongrel population. The farm colony is also characterized by free labor.

The tropical plantation colony, on the other hand, presents a marked contrast to the temperate farm colony in almost every respect. "The colonists are few in number, they do not contemplate an extended stay, and are represented preponderatingly by males: the racial unit is thus the individual, not the family."[15] In consequence, relations with native women produce a mixed-blood population. The motive of the colonists is to exploit the resources of the country for the home market, but since "vital conditions do not permit of the accomplishment of plantation labors at the hands of an unacclimatized race," laborers must be imported from other tropical regions if the natives cannot be coerced.[16] "Plantation colonies have regularly been the seats of wholesale enslavement," and the abolition of slavery leads only to various substitutes and subterfuges.[17]

The two opposing types of frontier agricultural organization grow out of differences in climate and in turn set up a different labor economy and a different social organization. This statement would seem to summarize Keller's explanation of the tropical or semi-tropical distribution of plantations. The limitations of the explanation as a theory of the plantation may be pointed out.

The plantation is a frontier institution which depends upon the development of trade, and the basis of trade lies in the exchangeable differences of peoples and areas. If the basis of trade is solely the difference in degree of industrial development between frontier and market, the exploitation of resources, agricultural or mineral, may be almost entirely independent of climatic conditions, and the plantation, or an organization similar to the plantation where resources other than agricultural are being exploited, may find its place anywhere along the frontiers of the world community. Exploitable mineral resources are notably distributed with little reference to climate. Colonial Pennsylvania had few plantations, if any, but the organization and operation of her mines resembled plantation organization and operation.[18] Plant life, however, unlike a mineral resource, is closely related to factors in the physical environment, and the natural distribution of the various forms of plant life is limited to the areas where the combination of environmental circumstances makes sustenance possible. Nevertheless, although the natural distribution of particular agricultural staples may be determined or limited by climate, the plantation form of organization and operation itself is not so determined. Eastern Prussian and the so-called Baltic Provinces seem to furnish a case in point.

The lowlands of Eastern Prussia have a very inhospitable climate. In some parts the period of vegetation lasts only from four to five months. Winters are probably more severe than in any other part of Germany. A good part of this territory was reconquered from Slavic peoples after the twelfth century and opened to German colonization. German merchants, knights, and monks were followed by peasant settlers with the heavy German plough. They drove out or subjugated the Slav, who with his light hook plough practiced only primitive methods of agriculture. The German peasant in turn gave way to a form of agriculture based upon large landed estates which began to export grain down the Elbe River to Hamburg and even to England. These estates seem to conform to the plantation pattern and have been maintained for over six hundred years.[19]

Difference in degree of industrial development as a basis for trade is, of course, a basis subject to change as the frontier comes of age. A more permanent basis of trade is found in the different agricultural products of different climates because differences in climate help to prevent the competition of tropical with temperate agriculture. Such a basis for trade is predominantly north and south in direction rather than east and west.[20] While exchange between temperate zone areas tends to become more and more specialized, exchange between the tropics and northern markets continues to be one of raw materials for consumers' goods.

The plantation organization of agricultural industry is largely concentrated in the tropical zone, not because of climate, but because tropical regions constitute the most important and most accessible frontier of the world community. They constitute a frontier where there are exploitable resources, mostly agricultural, that are nearer to consuming centers in terms of cost than are the vast areas of sparsely peopled lands capable of producing various kinds of agriculture in the temperate zones.[21] The reason the plantation predominates where it does is the necessity in those regions of securing a disciplined and dependable labor force. Where the native peoples are not sufficient in numbers or cannot be induced or coerced to supply the necessary labor, laborers are imported as indentured servants, as contract laborers, or as slaves. It is this rather than climate that gives its character to the plantations.

Not only are plantations entirely possible in temperate regions, but small farms characterize large areas in the tropics. Peasant proprietors may occupy the land with a self-sufficient agriculture; they may supply certain subsidiary requirements of adjacent plantations; they may produce the same staple as the plantations, selling to the latter or through the marketing organization which the plantations maintain; or they may market independently through cooperative associations. In the Southern part of the United States planters and small farmers have always maintained close relations. Between them, in the ante-bellum period, there was, according to Dodd, "no real economic competition or rivalry because of the sharp divergence of methods and products."[22] What we have to consider, in reality, is an economic and cultural complex in which either the small farm or the plantation pattern of social relations dominates.[23]

The Plantation as a Type of Settlement

Agriculture is, as Keller points out, the only important primary form of industrial organization common to colonies of all latitudes. Agriculture as compared with mining involves more permanent settlement, but the form which this settlement takes depends upon many factors, such as the availability of capital, labor, markets, transportation, and upon the nature and extent of exploitable resources. It also depends upon traditions, original purposes, and the extent to which these purposes can be realized or have to undergo modification in the actual process of settling and developing an area. When lands are occupied either by conquest, by migration, or by importation of labor, the tendency is to break up the familial and traditional organization of society and to create a proletarian labor class. But sometimes the soil is occupied by religious sects, having a social and moral order which remains intact. In such cases individualization and proletarianization may not take place.

Where migration or invasion brings together peoples of different stocks, different cultural traditions, different races and levels of culture, a period of turmoil ensues, after which a new society with new traditions and a new social order tends to grow up. What arises out of this turmoil, however, is a form of society no longer based on the family but on the territory occupied, a society no longer based on familial piety or custom but on common interests and competition. It is a more secular society.

How political institutions have their origin in turmoil and chance is exhibited in the rise of the *Polis* in ancient Greece. A passage from *The Rise of the Greek Epic* by Gilbert Murray describes the conditions under which this fortified city-state arose. It "is the sort of picture," he says, "that we can recover of the so-called Dark Age" when Greece and the whole Aegean area was the terminus of migrations from all parts of the Mediterranean and from the North.

It is a time, as Diodorus says, of "constant warpaths and uprootings of peoples;" a chaos in which an old civilization is shattered into fragments, its laws set at naught, and that intricate web of normal expectation which forms the very essence of human society torn so often and so utterly by continued disappointment that at least there ceases to be any normal expectation at all. For the fugitive settlers on the shores that were afterwards Ionia, . . . there were no tribal gods or tribal obligations left, because there were no tribes. There were no old laws because there was no one to administer or even to remember them; only such compulsions as the strongest power of the moment chose to enforce. Household and family life had disappeared, and all its innumerable ties with it. A man was now not living with a wife of his own race, but with a dangerous strange woman, of alien language and alien gods, a woman whose husband or father he had perhaps murdered—or, at best whom he had bought as a slave from the murderer. The old Aryan husbandman . . . had lived with his herds in a sort of familial connexion. He slew "his brother the ox" only under special stress or for definite religious reason, and he expected his women to weep when the slaying was performed. But now he had left his own herds far away. They had been devoured by enemies. And he lived on the beasts of strangers whom he robbed or held in servitude. He had left the graves of his fathers, the kindly ghosts of his own blood, who took food from his hand and loved him. He was surrounded by the graves of alien dead, strange ghosts whose names he knew not and who were beyond his power to control, whom he tried his best to placate with fear and aversion. One only concrete thing existed for him to make henceforth the centre of his allegiance, to supply the place of his old family hearth, his gods, his tribal customs and sanctities. It was a circuit wall of stones, a *Polis;* the wall which he and his fellows, men of diverse tongues and worships united by a tremendous need, had built up to be the one barrier between themselves and a world of enemies.[24]

What we have to recognize in this description of the chaotic conditions out of which the *Polis* arose and established order is that it likewise is a description, to some degree, of every frontier. In fact, this is just what we mean by frontier, a place of changing divisions of labor, a meeting point of cultures and of conflicting interests. If the American frontier was not the scene of "constant warpaths and up-rooting of peoples" to the same extent as Gilbert Murray pictures early Greece, the difference was merely one of degree. "[T]he western frontier worked irresistibly in unsettling population," says Turner, resulting in "that dominant individualism, working for good and evil," which is largely the significance of the frontier in American history.[25]

It is in this period of settlement, the period in which there is turmoil and disorder, that the plantation, like the *Polis,* arises as a means of establishing order, of restoring "the intricate web of normal expectations," and, unlike the *Polis,* of maintaining industrial and market relations. This double function of the plantation leads to enslavement or other forms of forced labor, to the importation of more individualized, or familyless labor, and to concentrated rather than dispersed settlement.

But the plantation is not the only means of bringing the resources of the frontier into new and higher uses. Where the invading people, the settlers, are able and willing to supply the labor, where the land occupied is used, in the first instance, in producing goods for consumption, there settlement may be achieved with free labor, with the labor of the settler himself. The settler brings his culture with him but does not impose it to any very great extent upon native peoples—presumably peoples with a more elementary type of culture. He supplants them. In such cases the settler tends to descend to the level of the culture of the natives whose lands he is invading. He becomes in America a trapper and a hunter. He may, as in the

Appalachian Mountains, revert to a type of society based on the clan and maintain a "blood feud." He is not included in the society based on the world market.[26]

These settlers become squatters, or illegal homesteaders. They pass outside the state and for a while, at least, become people without a state. They substitute experience for authority, and when they come to form political communities they regard them as being enacted by themselves; they do not regard themselves as coming from such communities. Competition and democracy become a tradition. Such a society evolved in the American Middle West. The Middle West had little to contribute to world commerce until its wheat fields were opened in the seventies of the past century, and Cyrus McCormick solved the labor problem by inventing the harvester, but by this time the Middle West had ceased to be a frontier.

Plantation settlement, or settlement where some type of large estate evolves to put land to "higher" uses and accommodate diverse cultural elements, involves, on the other hand, forms of forced labor and tends to acquire a feudal organization. Such a result was a fairly rapid one in South America following the Spanish and Portuguese invasions. "In a sense," says Professor Paul Reinsch, "the South American societies were born old. . . . The dominance of European ideas in their intellectual life, the importance of the city as a seat of civilization never allowed the pioneer feeling to gain the importance which it has held and still holds in our life. This backwoodsman of South America has not achieved the national and estimable position of our frontiersman."[27]

The Plantation and Labor

Some form of forced labor, such as slavery, seems to accompany the first organization of societies on a more complete territorial basis. Slavery arose in Egypt where the first city-states were established. It was an integral element in the organization of the city-state and of ancient civilization. Civilization in Egypt began with slavery.[28]

The conditions under which slavery tends to arise have been most systematically studied by Nieboer.[29] Limiting his investigation to ethnological evidence, Nieboer sought to discover where slavery exists among preliterate peoples, what kinds of societies have slaves and what kinds do not have them. In general, slavery is found to be more prevalent among fishing tribes than among hunting tribes, but not important among either. Neither is it important among pastoral nomadic tribes.[30]

Nieboer divided the agricultural societies into three groups. The first group includes those societies in which agriculture holds a subordinate place, since subsistence is derived mostly from hunting, fishing, and sources other than agriculture. Women only are occupied in tillage, and habitations are still frequently shifted. The tribes of the second group carry on agriculture to a considerable extent, but not to the exclusion of other means of subsistence; they have fixed habitations and lands are generally irrigated. In the third group agriculture is so relied upon that if fishing, hunting, or other sources of subsistence were entirely lacking, there would be little difference in the economic condition. On this level, lands are manured, rotation of crops is carried on, domestic animals are used, and agricultural products are exported. It appears that the increasing importance of agricultural production is a condition favorable to slavery, which prevailed among the commercial agricultural groups studied by Nieboer in the ratio of thirty-three to three.[31] The relative importance of agriculture is, of course, an index of the extent to which the principle of group organization is territorial rather than tribal.

The societies keeping slaves are found at each cultural level to live under conditions where subsistence is easy and not dependent upon capital, for the produce of unskilled labor can exceed the primary want of the laborer. Since everyone is able to provide for himself laborers do not voluntarily offer themselves. Such a society Nieboer calls a society of open resources.

> The most important result of our investigation is the division . . . of all peoples of the earth into peoples with *open,* and with *close* [*sic*] *resources.* Among the former labour is the principal factor of production, and a man who does not possess anything but his own strength and skill, is able to provide for himself independently of any capitalist or landlord. There may be capital which enhances the productiveness of labour, and particularly fertile or favorably situated grounds, the ownership of which gives great advantage; but a man can perfectly well do without these advantages. Among peoples with close resources it is otherwise. Here subsistence is dependent upon material resources of which there is only a limited supply, and which accordingly have all been appropriated. These resources can consist in capital, the supply of which is always limited; then those who own no capital are dependent on the capitalists. They can also consist in land. Such is the case when all land has been appropriated; then people destitute of land are dependent on the landowners.[32]

Among peoples with open resources slavery, serfdom, or modern substitutes for these are likely to exist, whereas among peoples with closed resources labor is "free." In the one case there are two masters running after one laborer, and the master who succeeds in getting him will probably take steps to hold him. In the latter case there are two laborers running after one master. Men must work for wages and in doing so must compete with their fellow workers, thus reducing wages to a low level, perhaps lower than the cost of keeping slaves. The difference between a situation where employers search for laborers and a situation where laborers look for jobs is just the difference between "forced" labor and "free" labor. Nieboer states the case as follows:

> Where subsistence depends on close resources, slaves may occasionally be kept, but slavery as an industrial system is not likely to exist. There are generally poor people who voluntarily offer themselves as labourers; therefore slavery, i.e., a system of compulsory labour, is not wanted. And even where there are no poor men, because all share in the close resources, the use of slaves cannot be great. Where there are practically unlimited resources, a man can, by increasing the number of his slaves, increase his income to any extent; but a man who owns a limited capital, or a limited quantity of land, can only employ a limited number of labourers. Moreover, as soon as in a country with close resources slaves are kept, they form a class destitute of capital, or land, as the case may be; therefore, even when they are set free, they will remain in the service of the rich, as they are unable to provide for themselves. The rich have no interest to keep the labourers in a slave-like state. It may even be their interest to set them free, either in order to deprive them of such rights over the land as they may have acquired in the course of time, or to bring about a determination of the wages of labour by the law of supply and demand, instead of by custom. They will thus, without any compulsion except that exercised by the automatic working of the social system, secure a larger share in the produce of labour than they get before by compulsion.[33]

Nieboer's theory, based upon his study of slavery among pre-literate peoples, indicates that slavery tends to arise in areas of open resources. In an area of open resources it is necessary to control and incidentally coerce labor in order to bring land into new and "higher" uses. Slave labor is likewise a form of capital with which surplus is produced to bring more land into new and higher uses.

Slavery, then, arises to solve a labor problem, but not labor in the family, and pre-literate society is organized on the basis of the family. Slavery does not flourish in pre-literate society, for the slave tends to be adopted into the tribe or family. Women, in patriarchal society, are the original industrial laborers, the first slaves, and males slaves captured from other tribes are made to do women's work.[34]

The Plantation as a Political Institution

Slavery as in industrial system does not impress itself on the form of society until the rise of the city and the city-state.[35] Theories of the origin of the state differ, but do not so much contradict as qualify and supplement one another. The essential problem in all of them is that of the transition of a society based upon the family and the sacred ties that bind it to the soil, to the mother country. Teggart points out that such a principle of organization as that resting upon kinship and embedded in tradition and religion would not, of course, be rationally and voluntarily given up.[36] Oppenheimer, following Gumplowicz, has systematically developed the thesis that the state historically arose with the conquest and forceful imposition of the authority of a nomadic people upon a sedentary and agricultural people.[37] The superior power and morale of the conquerors allow them to appropriate the energies of the conquered, who are bound to the soil and made slaves or serfs. Social and economic interaction within this state later tend to soften and limit the harshness of slavery or serfdom. The state originally begins its career as a large estate, and, Oppenheimer believes, the large estate may prove the last stronghold of the principle of exploitation which the state embodies.[38]

In short, as Oppenheimer conceives it, the state is also an institution of settlement for the purpose of putting land to new and higher uses. In the process, not duty, filial piety, but the common interests of individuals occupying a common territory become the controlling factor in social organization.[39]

The division of labor which exists between women (including slaves) and men in preliterate society takes the form of a division of classes with the introduction into this economy and society of subject races and peoples. Classes tend to assume the form of caste when the divisions of labor become more or less fixed. These classes and castes are, first, the laborers, and, second, the rulers (administrators). The rulers live in the city (the citadel) and the workers are on the land. The citadel becomes the guardian of the market place, where goods are exchanged. The rulers levy taxes and guarantee protection of traders as well as of agriculture and the agricultural laborers. The division of classes tends to become at the same time a division of labor between city and country.[40]

Conquest enlarges the boundaries of the territory which the city and the citadel dominate and protect. A conquering people, if there is conquest as there usually is, occupy the citadel and the conquered people on the soil merely change masters.[41]

When conquest takes place the conqueror becomes a King, not a tribal chief merely, and a feudal system is established. The essence of the feudal system is loyalty, not to a family, but to a leader or a territorial ruler who "possesses" the land. When the land is conquered the people who live on it are transferred with it to the rule of the conqueror. Nationalism grows up when, as in the case of Frederick the Great, the ruler arms the peasants and leads them to defend the soil they occupy. In the case of the Poles, the peasants were regarded merely as an indispensable appanage of the soil, which was owned and ruled by the nobles. Not until after the partition of Poland was the Polish peasant regarded as a *national* and encouraged to be patriotic.[42]

The gradual accumulation of incidents and variation of the sort which Oppenheimer describes in the evolution of the state leads, according to him, to the establishment of constitutional and public law, and of capitalism, in short, of the structure and artefacts that we call civilization. At the same time Oppenheimer states that relations between conquerors and conquered are softened and humanized. He apparently makes no distinction between these processes as they operate in the state and in the large estate. The large estate is regarded, apparently, as a sub-multiple of the state, or as the state in miniature. It is probable, nevertheless, that the process in which public and constitutional law develops accompanies the organization of society over wider areas, and is an impersonal process that operates in the larger state.[43] On the large estate the modification of relations is likely to be the result of more direct human and personal interaction between lord and peasantry. Oppenheimer describes this humanizing process as follows:

> Gradually more delicate and softer threads are woven into a net very thin as yet, but which, nevertheless, brings about more human relations than the customary arrangements of the division of spoils. Since the herdsmen no longer meet the peasant in combat only, they are likely now to grant a respectful request, or to remedy a well ground grievance. "The categorical imperative" of equity, "Do to others as you would have them do unto you," had heretofore ruled the herdsmen only in their dealings with their own tribesmen and kind. Now for the first time it begins to speak, shyly whispering in behalf of those who are alien to blood relationship. In this, we find the germ of that magnificent process of external amalgamation which, out of small hordes, has formed nations and unions of nations; and which, in the future is to give life to the concept of "humanity."[44]

The effect of more human relations, of patriotism for a fatherland, and of a religion which teaches brotherhood is to prepare the way for another culture which contains a new domestic principle of organization. To the extent that this takes place the state is maintained by evolving into its opposite—a homogeneous cultural group.

The Theory of the Plantation

The plantation is an incident in the conquest, settlement, and exploitation, politically and commercially, of an invaded country. It is an incident, if not of the origin, at least in the expansion of the state, that is, of the territorial organization of society. The plantation arises in areas of "open" resources as a means of controlling settlement, of preventing the dispersion of labor energy, and of concentrating it in the exploitation of the resources through the medium of a staple agricultural crop.

Under what conditions and circumstances do the resources of an area become open? Nieboer has answered these questions with respect to pre-literate groups. But we may suppose that at the level of pre-literate culture resources are open only in exceptional cases and do not remain open long. From Malthus to Carr-Saunders we have been impressed with the fact that the population problem among pre-literate peoples is as serious as it is among civilized nations. Certain customs such as infanticide and abortion are and have been extensively practiced by every so-called primitive group.[45] They are not practiced sporadically but strictly and regularly with much public feeling against their contravention; the elimination or prevention of excess population is a fact of constant operation. This leads to the important conclusion that, under a given state of resources and culture, there is an "optimum" population at which point numbers remain more or less stationary.[46]

It is evident that in a situation where population pressure against resources is so great that measures such as infanticide and abortion are sanctioned in the *mores,* the resources, at that particular level of culture, are closed, either economically or politically.

Resources are not open to a people who utilized the country up to the limit of their capital and technical means. Migration, however, may serve to introduce a group with new (superior) uses for the resources of the area and superior capital and methods for exploiting them. To the invading group the resources are open; closed to one group, they become open to another. Oppenheimer's herdsmen were men of superior power and morale compared with the sedentary people whom they conquered, but enslavement would scarcely have benefited the conquerors if the labor of one slave could produce no more than enough to barely sustain the slave, much less his master. It is necessary for the conqueror to introduce new methods to obtain a greater surplus as well as the power to enforce these methods upon the conquered. This probably is why Nieboer found slavery to be progressively more important in proportion as fixed habitations with higher agricultural methods became more important.

The invasion of a new area, however, is not always military. Indeed, in the present world community, the conquest of the city, and therefore of the country, is likely to be by a group which wins control of the market news and of the exchanges. Boundaries are enlarged by the migration of capital, speculative capital, as Alvin Johnson calls it, which invades the moral and cultural regions found "east of Suez."[47] These regions are to the holders of capital seeking investment, regions of open resources, regions where the mineral, agricultural, and labor resources of culturally inferior peoples may be made available to civilizations that, presumably, have superior uses for them.[48] If such capital is not able to evolve its own controls, however, it may be supported by military force.

The resources of an area, therefore, become open in the relationship between two cultures. Whatever form the invasion may take it is necessary, in understanding the outcome, to take into account the traditions and purposes of the invaders. Differences in traditions and purposes will mean differences in the form of politicization. Behind the agricultural missions of the Jesuits in Paraguay or of the Dominicans in the Philippines, for example, were religious traditions and purposes. Behind the Spanish plantations of the West Indies were purposes which did not contemplate colonization by white Spanish laborers but only by aristocrats who were to be granted lands and natives forthright. Negro slavery was not an evolution in the Spanish West Indies, as in Virginia, but was imported full-grown when the Indian population began to decline in numbers.

Where there is no outside market, the labor for exploiting the resources may be obtained by attaching the natives to the soil and exacting tribute from them in the form of ground rent. In this way the manor (*Grundherrschaft*) arises. If there is a large metropolitan market, resources may become open through the medium of the most profitable staple, whether agricultural or not. When it is an agricultural staple requiring routine labor in its production, the plantation (*Gutsherrschafter*) is likely to result.

Steady dependable labor is scarcely to be obtained from natives whose tribal culture resists the new habits unless their tribal culture can be broken down by force. But even where there is no moral objection to this by the invading group it may, nevertheless, be too expensive.[49] Unless it is broken down the native tribesman is valueless for plantation labor. In all plantation regions the native is described in similar terms—he is lazy, worthless and unreliable. An initial effort may be made to employ him but he is too free, too unaccustomed to the routine of steady industry, too near his family or tribe, too near familiar forest and streams, too difficult to force because too difficult to render dependent.[50]

For this reason "speculative capital," seeking to exploit the resources of an area through an agricultural staple, tends to rely upon imported labor which can be rendered dependent upon the employer and thereby held to steady industry. Recruited and imported labor is, by reason of the very conditions under which it is obtained and distributed, individuated labor rather than group labor. In the course of controlling it for industrial production the plantation assumes the character of a political institution, an institution whose principle of organization is territorial rather than tribal.

The extent to which some form of personal compulsion is necessary in the plantation exploitation of resources may be observed in different regions where such compulsion is abolished by outside interference. Such an opportunity is afforded by comparing the results of emancipation of slaves in the various regions of the British West Indies. In the densely populated island of Barbados, for example, where all land had long since been appropriated entirely, emancipation made little difference in the control of the former Negro slaves. They were simply employed at moderate wages not exceeding the former cost of maintenance. On the other hand, the abundance of fertile land not in plantations in Jamaica, Trinidad and British Guiana made it possible for the emancipated Negro to live without the necessity of regular work on the plantations, and their continual withdrawal into the interior left the planters with a steadily diminishing labor force. To meet the situation planters in these British colonies sought Hindu coolies from India who were introduced under terms of indenture for a fixed number of years.[51]

It is thus in close relation to the resources of a frontier area that the plantation arises. The factors in its formation are those which have to do with the traditions and purposes behind it, the development of its staple, the establishment of its market, and the controls which it introduces or evolves to concentrate and manage labor. In the process it becomes something like a state substituting common interest based upon locality and proximity for the ties of totem and blood.

The Plantation and Social Change

The plantation is an institution of settlement, a frontier institution. This means that the plantation is a transient institution and is involved in a cycle of change. Society always is in a process of change. Discussion of social change is concerned, first of all, with what it is that changes. If it is agreed that biological changes are not very important in social change, then we may say that the fundamental changes are in human relations, and in so far as these relations can be stated in their most abstract terms, that is, in terms of time and space, their changes may be subject to objective measurement or description comparable to the description of processes of change in a plant community. To state the case in terms of relations is to state it in terms of position and changes of position, i.e., of succession.[52]

Once a social order is established individual initiative, working within the limits which this order imposes, constantly modifies it in the interest of the individual. This aspect of succession is evolutionary. Institutions change at a different rate from the individuals who compose them. Individual habits change first; customs and forms change more slowly. Interest and attitudes of individuals constitute the content of customs and social forms. As the social order imposed by the institution gets established in custom and in the *mores,* it creates divergent attitudes of those who want to preserve it and humanize it and of those who seek to escape from, overthrow, or replace it with some other system. It is because one class regards it as a vested interest and another class is seeking to escape, that the institution has sufficient instability to permit change at all, or at least great and sudden change. Eventually the content bursts the traditional forms.[53]

The result is that institutions pass through a cycle of evolutionary changes, punctuated with catastrophic changes in the formal and traditional order. These catastrophic changes mark the periods of epochs in the natural history of institutions and societies. This series of catastrophic changes we may describe as institutional succession. The succession of formations in a plant community illustrates their nature.

Succession in a plant community, according to plant ecologists, begins with the invasion of a bare area by pioneer plants competent to survive under the extreme conditions presented by the bare areas. In adjusting to the extreme conditions, however, the pioneers effect a reaction upon the habitat itself, changing one or more of its factors in some degree. The reactions of the pioneer stage may produce conditions unfavorable to the survival of the pioneers but favorable to the invasion of new plant forms better able to utilize the area and so become dominant. The series of stages involved in plant succession proceed with the progressive interaction of habitat and invading life-forms which achieves final stabilization in a climax association, at which point the relationships between habitat and life-forms constitute a community closed to invasion by another dominant life-form.[54]

The natural history of the plantation community seems to follow much the same cycle of change. It, too, appears to arise in a "bare area," i.e., an area of open resources, after the way has been prepared by pioneering forms of economy. The plantation in turn prepares the way for further population invasions, a greater division of labor, and a more complex and resistant culture. The more or less extreme conditions of the frontier are modified as resources, even under the superior cultural uses of the newcomers, tend to become closed. Population increases, social relations become established, soil fertility diminishes, competing areas in the production of the staple crop arise, and changed market relations may lead to a more intensive agriculture. With these changes there seems to go a trend toward peasant proprietorship. Another change in market relations, the discovery and cultivation of a new and more profitable staple, a new invasion by foreign capital and superior methods may, however, interrupt the cycle at any point and start it over again.[55]

In more general terms, the territorial organization of society, the consequences of a process of politicization, involves an extension of the division of labor and the multiplication of artefacts—*structure*. It is the existence of this structure—including the division of labor—that we ordinarily call civilization.[56]

The structure of this secular society arises out of an order imposed upon peoples of diverse races and cultures who have experienced a "release," as Teggart puts it, from the bonds of old group relations. Normally, however, these individuated persons tend to be re-incorporated into a new set of relations and a new culture.

If civilizations, as Spengler says, grow up at the expense of existing cultures, they at the same time prepare the way for the development of new cultures.

But it happens, particularly in inter-racial situations, that there are factors which prevent the re-integration, or assimilation, of some individuals, at least at the same rate of speed. It is just in this connection that the biological fact of race mixture, where the races are culturally defined as virtually different species, has highly important consequences. The members of the various racial groups in such a situation get their conflicts settled by some more or less tolerable working arrangement with each other. But the appearance of the so-called "marginal man," who in this situation is a mixed-blood, introduces one whose lot it is to be permanently individuated because he is unable to incorporate himself in a system of relations, in a culture, in which there is no defined place for him. He is held, as it were, in a state of prolonged suspension overshadowing the system of relations in which others are at home. If he is unable or unwilling to "pass," or change himself, nothing remains except to agitate for change whereby the system of social relations may be ordered in such a way that he, too, can feel at home. The marginal man shows that the settlement of people, unlike the ecesis of plants, is finally a subjective fact.[57]

Virginia as a Typical Plantation Frontier

It is the purpose of this study to apply our theory of the plantation to a particular area. The plantation in Virginia affords an excellent opportunity for this purpose, for there the institution arose and seems to have passed through a complete cycle of change. With respect to the history of the plantation in this particular area it is necessary to distinguish between two historically related meanings embodied in the term "plantation."

To the people of Elizabethan England, and before, the word had reference to an assisted migration of people with a view to settlement. In this sense writers of the period spoke of the "plantation of Ireland" and the "plantation of America."[58] The acute need which England had long felt for timber gave rise to a good deal of propaganda designed to stimulate arboriculture, "the art of forming plantations of trees." The opening sentences of Bacon's essay "On Plantations" reveal this use of plantation to mean the transplantation of men like trees, both requiring capital that could afford to wait for returns upon the investment.

> Plantations are amongst ancient, primitive, and heroical works. When the world was young it begat more children; but now it is old it begets fewer; for I justly account new plantations to be the children of former kingdoms. I like a plantation in a pure soil; that is, where people are not displanted

to the end to plant in others. For else it is rather an extirpation than a plantation. Planting of countries is like planting of woods; for you must take account to lease almost twenty years profit, and expect your recompense in the end.[59]

The second historical meaning of the term, the one which is a commonplace in American tradition, is that of a relatively large estate with feudal characteristics producing agricultural products entirely for a market. The transition from a verbal to a substantive meaning occurred in all the English settlements along the American coast and took some interesting variations. Thus in Maine and New Hampshire it became the official designation of an unorganized and thinly settled division of a county. In the state of Delaware it is a legal term for an oyster-bed in which the oysters have been artificially planted. In Virginia and the West Indies, plantation became a large agricultural estate. Plantation as transplantation gave rise in these two areas to the plantation as an institution.

The genesis of the plantation of Virginia, and of the plantation in Virginia, is to be sought in the changes which occurred in what amounted to a revolutionary change in spacial relations between England and Virginia following the Period of the Discoveries. This revolution in distance, as measured in knowledge, time, and cost of transportation, shifted the ecological position of England from a place on the periphery of Europe to the center of an evolving world community. The same forces which brought about the shift in position were responsible for a certain amount of internal instability and relative overpopulation. Virginia, with great unused natural resources, was relatively unpopulated. Attempted relations projected along the lines of past English experience in dealing with overseas countries through trading factories proved inadequate. Unexpected conditions imposed modifications in original purposes and eventually changed these aims entirely.

The relationship between England and America constituted tidewater America into an area of open resources, but in the exploitation of these resources diminishing returns set in fairly early on account of physical difficulties in placing them on the market. In New England the soil was occupied by a religious group having a social and moral order which remained intact, so that extensive individualization did not take place. In the absence of a staple agricultural export a self-sufficing economy developed, characterized by the small farm, the town, and free labor.

Virginia, on the other hand, was settled by male adventurers who sought wealth and not homes in the New World. The discovery of tobacco proved to be a staple which could stand the expense of transportation to England and yield a profit. Moreover, the market was capable of great expansion. Control of production in its own interest by the Virginia Company failed because of the difficulty of long distance communication, and the resident planter arose as a responsible agent for the employment of English capital in the production of tobacco.

The need for labor was now transferred from Company to planter. Unable to coerce the tribal Indian, the planter became an immigration agent importing white indentured servants whose wages he virtually paid in advance. His method of controlling such labor was necessarily by lengthening the period of servitude, by corporal punishment, or both. Negroes introduced into the colony originally assumed the customary status of indentured servants, but an accumulation of incidents led to slavery as a customary and, finally, a legal relation. The slave trade developed from the economic and legal relation and introduced the harshest features into the system only to be modified and humanized again by the establishment of new customary relations. While Negro slavery was in process of evolution a white planter class developed simultaneously.

The plantation which arose as a political institution *in* settlement, both in Virginia and the West Indies, now became an institution *for* settlement. The increasing demand for cotton, the industrial revolution in England, the invention of the cotton gin in America, the surplus capital in the form of slaves, the soil and climatic resources of the South, and relatively easy transportation by means of rivers and the Gulf of Mexico all combined to constitute "the South" into an area of open resources. The plantation moved southwest with the advance of the frontier, exhibiting its greatest activity along the frontier and undergoing changes as the frontier moved on. Before emancipation the presence of the free Negro was an index of these changes.

The conditions under which the plantation arose in Virginia have perhaps undergone the greatest changes of all the plantation communities of the United States, for the frontier there has more completely passed away, and with it the plantation. Virginia, today, is predominantly a state of small farms.

2

ℳ

The Metropolis and the Plantation

The Revolution in Distance

"The secret of environmental control," says McKenzie, "lies in the ability to conquer distance."[1] The series of changes which began in the period of the Great Discoveries which led to, and followed, the Industrial Revolution resulted in changing what were once only geographical relations into an ever widening circle of economic and political relations. This series of changes might be described in general terms as a revolution in distance, since it greatly increased control over a larger part of the physical world by making the resources of every part more accessible to every other part. These changes laid the basis for a knitting and integrating process which has achieved our present world community.

The first phase of the revolution in distance was the series of discoveries which substituted the Atlantic Ocean for the Baltic and Mediterranean Seas. It was a veritable *geographical revolution* since it opened up new countries for conquest and trade.[2] First returns from the New World were gold and sliver when, after 1520, the Spaniards pillaged Aztec and Inca treasures and then developed a steady supply by working the mines of Peru, Bolivia and Mexico. The effect of this impact of precious metals was to "alter profoundly the entire structure of economic life on the continent of Europe."[3] Among other changes involved in the *financial revolution* made possible by the exploitation of the mines of America was the substitution of a money economy for what had been largely a barter economy.[4] A money economy made possible a mobile capital on a vastly enlarged scale without which extensive combinations for commerce and industry and overseas expansion would scarcely have been possible.

The financial revolution in turn contributed to social disorganization and further changes. While the theory that the substitution of a money economy for barter led to the commutation of personal work by the serf for the lord to money payments, which in turn gave the serf his freedom, has been shown by Nieboer[5] to be untenable, yet it undoubtedly contributed to the so-called *agrarian revolution* by facilitating the exchange of goods. Far-reaching changes took place when the lord began to play an active rather than a passive role in increasing his net revenues from the land. The resulting Enclosure Movement forced the uprooting of thousands of peasants from the soil and made them a labor market for mine, factory, and plantation.[6] Although the standard of living in England was rising, the presence of a large begging army of poor gave rise to the opinion that England was over-populated,[7] and this conviction was fed by the contemplation of inviting and almost unpopulated lands across the Atlantic. The peopling of a new continent was a great revolution in the *distribution of population* in both the New World and the Old. One of the significant things about this redistribution was that the new population in the New World did not merely take the place of displaced natives, but it constituted a group which consistently maintained relations across the Atlantic with the Old World. This is a highly significant fact about the plantations. They constituted the Atlantic into an area of interaction, an interaction which continually changed the characters of all countries connected with it and gave them a common European culture. It created what Ramsay Traquair calls the Commonwealth of the Atlantic.[8]

Commercial relations with the new populations across the Atlantic were at first not so significant for any increase in the amount of trade as they were for changes in the highways and goods of commerce that came with them. These changes in the routes of commerce constituted the *commercial revolution*.[9] Without any change in the specific goods of commerce, however, the commercial revolution might never have led to those mechanical inventions and large-scale manufacturing methods whereby the factory supplanted home industry and the guilds. In increasing amounts certain goods from overseas came to England which required further processing before final consumption. Chief among these goods was cotton, and it is not without much significance in this connection that the inventions commonly associated with the *industrial revolution* in England were those for the manufacture of cotton textiles.[10] The industrial revolution in England and the cotton plantation in the South were part of the same set of facts.

The position of England off the western coast of the European continent proved a vantage point of great strength in the control of the network of relations over the area of the Atlantic. This advantage of position was augmented by the fact that, as an island in the midst of maritime enterprise, she was favored by cheaper transportation per unit area of land than continental nations. Ships and shipping became an important part of the English culture.[11] Throughout all the changes involved in what we have called the revolution in distance, England remained disorganized and unstable enough to respond sensitively and vigorously to overseas developments.

The trading company emerged out of these changes as a powerful agency of empire. Outside of England, omitting France, Spain, and Portugal, the world was practically mapped out and allotted with monopoly rights for the English trade to the various companies.[12] It is worth noting that these companies grew out of very much the same conditions which later provided some of them with the human material for overseas plantations. These conditions had to do with loosening economic and social bonds out of which individuals emerged from old group relationships free to seek status and security in new channels. The original tendency toward associations was due to the insecurity of medieval trade. The trading company was, however, only one way in which men were being realigned into new group associations. Scores of social movements strove to recapture the old values of manorial England which trade had partly destroyed. The Puritan movement and others appeared, and some sought the good life across the Atlantic.

This was a period when forces over which the individual Englishman possessed little control were breaking down the old routine of settled life. He was threatened with losing, if he had not already lost, the "place" in a system of relations which he had inherited from his fathers and which gave him security and mental ease. Now if he was not actually poverty-stricken, as a great many of his fellow-countrymen were, he was thrown into the company of strangers who were not likely to accord him the rights and privileges which he had traditionally enjoyed on a manor. Other rude fellows were competing for his place, and he himself aspired to a higher status. There were more goods of commerce in the new order, but the competition for them was keener and the enjoyment of them less satisfying. What the individual Englishman of the time probably wanted was what men always want, a social environment in which he might have a satisfying conception of himself.

A law of 1536, the first of the English Poor Laws, attempted to restore the lost social equilibrium by reestablishing geographical and social fixity. Later laws established several institutions for dealing with the poor, among them the workhouse and the allowance system.

If the emphasis of the Poor Laws was upon effecting poor relief by achieving stabilization and fixation in the home community, plantation, in so far as it, too, was a philosophy and technique of poor-relief, would solve the problem in an opposite manner, by transporting the poor and insecure to overseas territories. Few propagandists for plantations failed to point out what they considered to be a surplus population in England whose condition they felt would be bettered by the migration. "A Plantation," wrote William Penn, "seems a fit place for those Ingenious Spirits that being low in the World, are much clogg'd and oppress'd about a livlyhood, for the means of subsisting being easy there, they may have time and opportunity to gratify their inclinations."[13]

But men do not participate in social movements without some definition of goals and purposes, even if somewhat vague. Not a minor but a major role in the

English plantation of America must be given to the New World's function in providing an objective, an organizing principle of restless impulses at a time when an old agrarian order that had furnished security was breaking down. One expression of the growing conception of America in ideal terms was Thomas More's *Utopia*. "His vivid imagination first brought together in fruitful union the world of Plato and the world of Columbus and Cabot."[14] On his mythical island in the West (it was not then known that America was a continent) More painted his picture of an ideal society, "a new community founded on peace, goodwill and equity"— and security. Such a society became all the more ideal when contrasted with the disorganization in England incident to the craze for turning tillage into pasturage, wherefore sheep "devour whole fields, houses, and cities."

In short, for a number of reasons there were men ready to go, and agencies to effect their transportation to the New World were not lacking. If there had existed no ocean between England and this promised land undoubtedly an unorganized rush of individuals to America, similar to the gold rush to California, would have occurred. But the presence of this barrier, with the equipment and expense required to surmount it, necessitated a high degree of prior organization. It is a fact of present-day observation that whereas rural individuals in large number will move to the city without urging, migration from the city to the country seems to require much propaganda and tends to assume the form of a social movement. Plantation was just such a migration, which was at the same time a social movement, and as such required organization and some definition of purposes.

As a migration and a social movement, plantation was originally a reaching out for Utopia. Some wished for the utopia that would restore the traditional sacred values of England's lost golden age, others for a secular utopia of wealth, property, adventure, and fame. Representatives of one group went with their families to New England; representatives of the other group went without their women to Virginia. The planting of people across the Atlantic affected a revolution in the distribution of population and was a part of the greater revolution in distance whereby men progressively began to get control over more of the world's resources and to make them available to each other through commerce and manufacture. "If I had to choose a single characteristic of Romance as the most noteworthy," says Sir Walter Raleigh, the essayist, "I think I should choose Distance, and should call Romance the Magic of Distance."[15]

The Trading Factory

The revolution in distance, the uneven shrinkage of the world community in terms of time and cost of transportation and communication, has presented itself to geographers as a problem in improperly conceiving and representing present geographical "realities," that is, new spacial relations that are out of line with our traditional conceptions. One of these geographers is H. J. Mackinder whose book, *Democratic Ideals and Reality,* deals with the new scale of distances that have been ushered in by new forms of transportation and communication and by the redistribution of population.[16] The ocean, for instance, is one and has been throughout history, but its tremendous expanse led men to conceive the water surface of the earth as five great oceans. The revolution in distance, however, has made these arbitrarily designated sub-oceans unreal for the present reality is the "world ocean."

For the same reason that the unified ocean has become a practical reality "the joint continent of Europe, Asia, and Africa is effectively, and not merely theoretically, an island"—the "World Island."[17] To the ancient Greeks this land mass was the "world." With better transportation and knowledge it was reduced to a group of continents. Now under modern transportation and communication it is no larger than some islands formerly were. With its smaller adjacent islands the "World Island" occupies only one-sixth of the earth's surface but holds fifteen-sixteenths of its population. Within the World Island four-fifths of the population live in two great "Coastlands," a western Coastland consisting of most of Europe, and an eastern Coastland comprising India, China, [the] Malay Archipelago, and Japan. In area the Coastlands together measure only one-fifth of the entire World-Island. The remaining area constitutes two great "Heartlands" which, even under modern conditions, do not have access to ocean transportation. But our interest here is in the Coastlands.

Mackinder does not go very far in analyzing the differences between the two Coastlands, but there are and long have been fundamental differences. Brunhes and Vallaux point out that the population of the European Coastland is concentrated in metropolitan cities (*concentration active*) whereas the Oriental Coastland is a region of rural population concentration (*concentration passive*).[18] India, in spite of her dense population has few large cities but the regions of heavy density in the European Coastland are metropolitan regions and are possible because specialization of function enables them to reach out and make the rest of the world a hinterland.

The designation of the two types of population concentration as "active" and "passive" suggests a fundamental contrast between the two Coastlands. During the middle ages the European Coastland was probably almost as "passive" as the Eastern Coastland, but it gradually became "active" in the process of building up relations between its own parts and of entering into relations with the East for which it took the initiative. It was in the accomplishment of these relations that the revolution in distance had its beginnings. They were originally, of course, trade relations, and both in Europe and the Far East the trading factory[19] arose as a means of making and maintaining them.

In his study of the state, Oppenheimer gives a general explanation of how trade relations between coastal peoples arose. Sea nomads, like land nomads, are accustomed to prey upon settled coast peoples and seize their property as booty. But even more than the land nomads they cannot well do without markets, for much of the loot which the victors have taken in their raids consists of property which is more valuable for trade than for immediate consumption. Just as the professional thief needs his "fence," the Viking, or sea nomad, needs harbor markets in order to dispose of his property. Many of the cities around the Mediterranean, Oppenheimer thinks, were originally places for trade in goods seized by pirates.

> The harbor markets developed from probably two general types: they grew up either as piratical fortresses directly and intentionally placed in hostile territories, or else as "merchant colonies" based on treaty rights in the harbors of foreign primitive or developed feudal states.[20]

From some of the harbor markets of the first type developed territorial states in which the Vikings became a landed nobility. "Where, however, the Vikings did not meet peaceable peasants, but feudal states in the primitive stage, willing to fight, they offered and accepted terms of peace and settled down as colonies of merchants."

> We know of such cases from every part of the world. . . . To take the instances with which Germans are most conversant, there are the settlements of North German merchants in countries along the German ocean and the Baltic Sea, the German Steel Yard in London, the Hansa in Sweden and Norway, on the Island of Schoen, and in Russia, at Novgorod. In Vilna, the capital of the Grand Duke of Lithuania, there was such a colony; and the Fondaco del Tedeschi, in Venice is another example of a similar institution. The strangers in nearly every instance settle down as a compact mass, subject to their own laws and their own jurisdiction. They often acquire great political influence, sometimes extending to dominion over the state.[21]

It is characteristic of these merchant colonies, of "factories" as they came to be called, that they tended to extend their political influence into complete domination. They were enabled to do this by developing strength among their own numbers or by the aid of the home state. Through her factories in India, England was enabled to acquire dominion over that country. On the other hand, the German Hanseatic merchants in England remained powerless aliens and eventually were expelled from the island. At any rate, the factory is seen as an institution which everywhere arises under similar conditions, where maritime people of one culture seek to dispose of their commercial capital, however obtained, for the wares of an oversea people of another culture. Western relations with modern China exhibit essentially the same phenomena.

In connection with our present study there seems to be three reasons why we should consider the trading factory. In the first place, it was that part of the original pattern of English overseas relations which had to do with actual migration and settlement. In numbers this movement of population was relatively small, since the volume of trade and the means of its transport required the maintenance of only a few "factors" in a factory, who had nothing to do with the production of the commodities in which they dealt. This was the work of the native population among whom they lived; very rarely did the factory go beyond its commercial purposes; when it did so its character as a factory changed.[22] Again, the factory directs attention to the agency which formed the active connecting link between overseas countries and which in Genoa, Holland, France and England developed into the regulated or joint-stock trading company. The trading company, in turn, was nurtured in the countries that were getting cities, an "active" concentration of population, and represented their reaching out for larger supplying areas or hinterlands for which they were in competition and conflict with each other. The factory was the overseas instrument of these companies, their point of commercial contact with peoples of another culture in a foreign country. At this point of contact the institution was crystallized.

Finally, the factory and the trading company contributed to the secularization of those countries that maintained them. From her established relations with old and well populated Coastlands in Europe and Asia, as well as from internal developments connected with the rise of cities, England, as one of these countries, gained commercial experience exceeding that of her rivals and became a "nation of shop-keepers." She not only gained skilled commercial experience, but her merchants became educated with respect to the value of the resources of other countries which England might turn to superior uses. This experience was deposited in her tradition and culture and served to define her original purposes with regard to America. Here was another country with resources and wares to be traded for as the resources and wares of the Baltic, the Levant, and the Far East were traded for.

England's overseas trade had its more significant beginnings with the Merchants of the Staple, those merchants who dealt, for the most part, in English wool, and who, with the king, were much concerned to regulate and standardize it—to make it "staple," both for purposes of trade and for taxation. Their organization was at best a very loose one and was displaced in the sixteenth century by the Merchant Adventurers of England. The overseas trade of the Staplers was largely confined to staple towns in the Low Countries across the English Channel. The Merchants of the Staple seem originally to have been foreign merchants, "merchant strangers," for England was itself a frontier region where continental merchants went for wool. England's development as the center of the new economic organism that was in process of formation is seen in the establishment of staple towns in England itself and the rise of an English merchant class. At the close of their career

> the Staplers became the home merchants, mainly concerned with the export of the raw product, wool. The Merchant Adventurers became the English merchants domiciled or sojourning across the seas, in foreign parts though near home; they were concerned with importing into the cities and lands wherein they planted themselves, no wool grown in England, but cloth made in England from English wool.[23]

The earliest charter granted to the Merchant Adventurers was not to give a trade monopoly, unlike later charters, but was simply to enable Englishmen living abroad to govern themselves. English merchants in the Low Countries of Europe who had secured factory concessions from the cities in which they resided probably organized themselves for their own government and trade and then applied to their king to recognize their organization. The result was a charter which made their proceedings legal. It is significant that the central control of this *imperium in imperio* was not in England but across the sea at Bruges, Antwerp, Middleburg, Hamburg, or wherever the main mart was established at the time. The governing body to whom the members swore allegiance was composed of a governor and twenty-four assistants all domiciled across the sea and elected by the members of the fellowship residing in the main mart. To these overseas officials all members were subordinated, even those residing in England. In the evolution of the English trading company, therefore, the trading factory was prior in point of time and importance to centralized control in the home country. The centralization of control in England never became an official fact in the case of the Merchant Adventurers, although the number and influence of its London merchant membership residing abroad led to frequent protests against the domination of London.

In 1404 Henry IV granted a charter to the Eastland Company, which undertook to plant factories in the countries bordering the Baltic in the face of opposition from the Hanseatic League. The importance of this short-lived company is that it, for the first time, centered control officially and exclusively in the city of

London. Centralized control in England instead of peripheral control from abroad was, from the Eastlanders on, to characterize all English trading companies.

The Eastlander carried English trade farther afield than the Merchant Adventurers, but trade was still confined to Europe. But Englishmen, in common with other Europeans, were anxious to obtain the gems, silk, drugs, perfumes, and spices of Cathay, the trade of which was monopolized by the Portuguese. How to reach the Far East without coming in conflict with the Turks in the Levant or the Portuguese along the Cape of Good Hope route was for the English a difficult problem. Many believed that a northeast route along the northern shore of Asia was possible. Accordingly in 1553 a company was chartered for the purpose of finding such a route, but experience taught it that no profitable trade was to be found in those parts except in the dominions of the Czar, with the result that the company became the Russian or Muscovy Company and gave up its extensive ambitions. But it discovered the possibilities of profitable trade in the heart of Asia from the ports of the Levant. This led to the chartering of the Levant Company about 1575 to tap these trade sources. The Venice Company, organized in 1583, contemplated gaining the trade of the Mediterranean and fixing its control in English hands. Later the Levant, or Turkey Company, and the Venice Company were amalgamated. The Levant Company was the first English Company to gain concessions and establish factories in non-Christian lands. The experience gained in dealing with Mohammedan peoples stood the English in good stead when later through the East India Company they finally reached the Far East.

The failure of the Muscovy Company emphasized the necessity of a larger and stronger organization to compete with the Dutch and Portuguese if the English were to deal directly with the East Indies. In 1599 a company of London merchants subscribed a stock of over £30,000 in the organization of the East India Company, for which a charter was secured the following year and the first voyage sent out "for the honour of the country and the advancement of the trade of merchandise." It became the first important English joint-stock company.[24] The joint-stock principle of control and organization was incidentally evolved in connection with the East India Company out of the practice of sub-groups of merchants banding themselves together in order to finance particular voyages to distant India. Beginning with Surat a number of factories were established in India, and the English adventure in the Far East was under way.

English factories overseas and English companies engaged in carrying on a trade between them and English ports laid the basis for the first British empire. From 1598 to 1763 this was, according to Mrs. Knowles, an Empire of Outposts consisting of islands, trading factories in Europe, India, and West Africa, and a stretch of sparsely inhabited coast-line in North America. Accessibility to ocean vessels was the characteristic feature of this empire, for it was essentially a trader's empire. There was little expansion into the interior of the overseas countries traded with.[25]

The Empire of Outposts (and the outposts were principally trading factories) was the general result of efforts in which the English people took a leading part to establish relations between the two great Coastlands of the World Island. In the course of establishing such relations the European Coastland developed an "active" concentration of population in cities seeking and competing with each other for larger hinterlands. The New World was discovered as an incident in the development of relations between the two Coastlands. In an interesting article Charles Redway Dryer shows that historically the New World, extending from one polar ocean to the other and lying between the two Coastlands of the World Island, proved an effective barrier to direct east and west communication between them. Magellan and Drake finally found the Cape Horn route around the barrier, but the long search for a northwest passage was a failure. In effect the New World, with the Arctic ice fields and the Antarctic land and ice, is, as it appears on such a map as Mollweide's projection, a ring—the "World Ring"—which almost surrounds and encloses the World Island.[26]

In the efforts of Spanish, Portuguese, French, and English to pierce the barrier of the "World Ring"—the two Americas—something was learned about it. The Spaniards discovered gold and silver and this raised the hopes of the other countries. The English sought gold and new sources of trade. They did not find either. For, unlike India, the New World, and especially that part available to the English, was very thinly populated with a people whose economic wants and trading habits were relatively little developed.[27] These facts had important consequences. They meant, for one thing, that the exploitation of the New World would require methods and techniques different from those used in the Far East. An extensive commerce was not possible in America because there were only a few natives to trade with, and these had no manufactured or semi-manufactured goods to be traded for. Neither did they have any conception of the value of the raw materials of the country which Europeans wanted and for which they were ready to barter. The pelts and skins secured by Indian hunters were valuable, but where there were possibilities of other resources, in the acquisition of which the Indians had no experience, the trading factory proved insufficient.

The exploitation of these resources required a settled population who knew their value and who were willing to work, or could be forced to work, in order to make them available to the market. Under these circumstances the commercial factory in Virginia evolved into the industrial plantation, but not without a great deal of trial and error in which all the experience of the past was brought to bear upon the new experiment of bringing the resources of North America into English commerce. To this end Richard Hakluyt set about making a systematic collection of records and papers concerning English maritime enterprise in all directions; in his own words, to "make diligent inquiries of such things as might yield light unto our western discoveries in America." Out of his efforts came his monumental work, *The Principal Navigations, [Voyages,] Traffiques, and Discoveries of the English Nation*.[28] The English Plantation in America was to be a development, in the words of Carlyle, of an "old habit or method, already found fruitful, into new growth for the new need."[29]

His Majesty's Plantations

The transition from trading to industrial motives in English relations with America is especially well illustrated in the series of documents contained in the eighth volume of Hakluyt's *English Voyages*. Thus Sir George Peckham wrote in 1583 that the principal purpose of Englishmen in America was to dwell peaceably among the natives "and to trade and traffique with them for their own commoditie, without molesting or grieving them any way."[30] Of the West Indies he had heard "that the people in those parts are easily reduced to civilitie both in manners and garments." If this was true "what vent for our English clothes will thereby ensue."[31] Captain Christopher Carleill in the same year addressed a brief and businesslike letter to the merchants of the Muscovy Company arguing the uncertainty of English foreign markets and New World possibilities. He expected that the natives of America would

> daily little by little forsake their barbarous and savage living, and growe to such order and civlitie with us, as there may be well expected from thence no lesse quantitie and diversities of merchandize than is now had out of Dutchland, Italie, France or Spaine.[32]

These writers, and many others, specified in some detail the nature of the commodities which might be obtained from America. Carleill listed fish, furs, hides, naval stores—which then came from Russia—and thought well of the prospects for wine, olives, wax, honey, and salt. In addition to these Peckham included silks, dyes, palm wine, and precious stones. "It is hereby intended," he said, "that these commodities in this abundant manner, are not to be gathered from thence, without planting and settling there."[33]

Carleill and Peckham apparently did not intend that the planters should engage in the actual labor of producing needed commodities, but should obtain them by trading, discovery, and treasure-hunting. It was Thomas Heriot, one who had actually visited Virginia and had seen for himself as the others had not, who was one of the first to give to systematic cultivation by the settlers a prominent place among the motives for peopling the New World. In his *A Report of Virginia*, Heriot divided the products of the country into two parts: commodities for the sustenance of the plantation, and commodities for merchantable export.[34]

By the time he wrote his Epistle Dedicatory to Sir Walter Raleigh in 1587, prefixed to his translation of a history of Florida written in French, Hakluyt also had come to consider actual cultivation of needed agricultural commodities to be a motive for plantation. He argued that the Spanish and Portuguese had pursued such methods with great profit and Englishmen might well follow their example.[35]

The motive of plantation for purposes of agricultural production rather than for trade was thus reached through reflection by writers long before English merchants and actual planters had reached through experience the same conclusion. We shall see later how the adventurers and planters of the Virginia Company were forced through circumstances to abandon hopes for a large trade and to rely upon cultivation.

For something like a quarter of a century before the chartering of the East India Company individuals had been experimenting with a new kind of undertaking designed to populate, or "plant," the New World. The efforts of these individuals, among whom Sir Humphrey Gilbert and Sir Walter Raleigh were the most outstanding, ended in failure but paved the way for the more successful companies. We have seen how Carleill attempted to persuade the Muscovy Company to turn its attention to America.

> But when this of America shall have been haunted and practiced thirty yeeres to an ende, [he argued,] as the other [the Russian trade] hath been, I doubt not by God's grace, that for the tenne shippes that are now commonly employed once the yeere into Moscovia, there shall in this voyage twise tenne be employed well, twise the yeere at the least.[36]

But the old companies did not change their course from known channels of trade to unknown ones. Rather new companies arose, organized largely by individual members of the old companies, which were specifically authorized to transport and plant settlers in overseas territories. Here is where they began to differ from the trading companies.

> The English trade with other nations of western Europe and even with Russia and the Levant involved most prominently the establishment of international relations, whether directly through the national governments or indirectly through the medium of commercial corporations; the trade with India and Africa was to involve rather the absorption of the governments of

the foreign peoples; but the colonial trade involved an extension of the national government of England over bodies of its own subjects. The colonial commerce did not consist merely in exchanging English products for the goods produced by foreigners through their development of the natural resources of their land, but much more largely in the primary production of goods by direct development of natural resources; the purposes had to be accomplished by actually settling the land with English colonists. Moreover, the tracts of land colonized were manifestly part of the domain of England and not of foreigners.[37]

While the efforts of each of these new "planting" companies encountered different problems and had different consequences in the settlements which they planted from Newfoundland to the West Indies, nevertheless the activities which they promoted, when viewed as a whole, reveal the fundamental sense in which Englishmen used the term plantation when they had reference to a certain form of migration. For they were all activities directed toward the settlement of population which involved a settlement of capital as well.[38] Considerable capital accumulation and investment were required to sustain an overseas migration and settlement during a more or less lengthy period of enforced unproductiveness. Overseas settlement became a problem of capitalistic "planting" of men, of managing their labor, and of waiting for returns. The capital required for transplanting the sort of English laborer who was willing to go was likely to be entirely beyond his reach. It was the planting company or organization that provided it.

This is plantation as the Englishman of the period, standing at the point of its origin in England, viewed it. In America, along a thousand miles of Atlantic coastline, prearrangements, intentions and traditions underwent profound modifications in each settlement. In each company's plantation there was originally an absence of private property because the planters were factors or tenants on the lands of the company. As in the trading factory, the government of the plantation was expected to be in terms of the interest of the company. The economic unity upon which this company government rested began to disintegrate with the appearance of private property in land, for economic enterprise was then no longer under unified control. In such widely separated settlements as New Plymouth, New Netherland, and Virginia, the introduction of private property in land initiated similar changes. In Virginia the private ownership of property began in 1616 and company ownership ended in 1619. In New Plymouth the period of proprietary economic control was a brief one, and private ownership was established in 1623. The Dutch West India Company offered private plantations to free planters in New Netherland in 1629.

Another unexpected departure from original intentions, and one which inevitably followed from the introduction in each settlement of private property in land, was the differentiation of local government from general government.

The governments of the early plantation colonies had in them the elements of both local and general control, managing as they did the actual interests of single small settlements and yet holding the powers necessary for governing the whole region in which a settlement lay. At first these colonial governments were essentially local in nature. When settlements multiplied, the extensive powers of the several executives, which had been possessed from the beginning, were utilized to enforce political unity. The change brought no break in the sequence of colonial administration. The word "colony" merely took on a broader meaning than before, while "plantation" remained what it had been, a local community subject to colonial government. The plantation type is therefore the ancestor of the older colonial and state governments by direct derivation.

But the plantation type begins not only the development of colonial government but that of local government as well, for as agricultural settlements multiplied beyond the first simple establishments, the various features of the plantation type reappeared in the new communities. Usually these features were more or less modified in their extent and completeness, but still they were characteristic, and their presence marks off broadly a certain large group of local governments as radically different in nature from the local communities of the present time. In this group are included the privileged plantations of Virginia, the manors of several colonies, the patroonships of New Netherland and many of the New England towns. The kinship of these places to the plantation type is plain. They were based upon agricultural organization. There were in each a measure of economic unity, a combination of jurisdiction with powers of proprietorship, and some use of civil administration for economic ends. This group of modified forms includes also such settlements as that of the Massachusetts Bay Company, which, like early Jamestown, was both plantation and colony, but which was not of the pure plantation type. An evolution went on in these modified forms in much the same way as it had in the first colonial plantations. Sometimes the course of events stripped away the jurisdictional side of a settlement and allowed it to fall back into a mere personal estate, but more often the economic side was given up and the community developed into a political entity with only political powers.[39]

Scisco in the above quotation links up in a very interesting way the plantation large estate in Virginia, the manor in Maryland, the patroon in New Netherland,

and the town in New England as results of a similar process of development under the different conditions of each colony. The kinship of the New England town and the Virginia plantation, usually contrasted by the interpreters of American institutions as two polar extremes, is here shown. The New England plantation, such as Salem plantation, became a town,[40] while Virginia plantation became a large agricultural estate in a colony which remained predominantly rural.

The English settlements in America, from New England to Virginia, had sprung out of the economic and commercial traditions of England. But the necessity for adaptation to the ruder requirements of the frontier produced in each a reversion from a partly artificialized environment to a natural one. A certain amount of acculturation took place from the native Indians to the colonists; to some extent the colonists became Indianized. The very forms of social and economic organization, according to Sumner, were adaptations on lower levels.

> No civilized people have ever had so little civil organization as the colonial [New England] towns early in their settlement; there was little division of labor, scarcely any civil organization at all, and very little common action. Each town was at the same time a land company and an ecclesiastical body, and its organization under each of these heads was more developed than in its civil or political aspect. The methods of managing the affairs of a land company or a congregation were those of the town as a civil body also and the different forms of organization were not kept distinct. The administration of justice shows the confusion most distinctly; all common interests were dealt with by the one common body without distinction or classification; and as committees for executing the decisions of the body were the most obvious and convenient device for executive and administrative purposes, we find that device repeated with on slight variations. . . .
>
> In the South, where the plantation system existed, not even these nuclei of social organization were formed. Thus the whole of this country, until the beginning of the eighteenth century, presented the picture of the loosest and most scattered human society which is consistent with civilization at all, and there were not lacking phenomena of a positive decline of civilization and gravitation toward the life of the Indians. Political organization scarcely existed and civil organization was but slight. . . .[41]

This picture of colonial America as the "loosest and most scattered human society which is consistent with civilization at all" is in accord with the facts. However, the plantation system of the South, which Sumner seems inclined to pass over, not only evolved out of the conditions which he describes and established order, but it maintained the first and, until the last century, the most important specialized industry in North America. The plantation's conditions of economic dependence on account of its specialization in staple crops, such as tobacco and cotton, helped to create divisions of labor for non-plantation settlements and thereby helped integrate all the American settlements. Only in the South and in the West Indies were there large populations that did not produce their own food requirements. It was largely because of this fact that New England, whose forests failed to provide the expected staple export, was able to make the fish and provisions which she sold to the plantation settlements balance her imports.[42] In the same way, two centuries later, the wheat and meat of the West found a market in the Southern plantations.

Although pressed closely against the soil and not without its cultural variations in consequence of its rural life, the plantation estate helped to extend and maintain a little of the urban and artificialized environment of England on the New World frontier. It was partly through the plantation that the economic traditions of the trading factory were perpetuated and applied in the new situation. Without it, the "positive decline of civilization" in the South, of which Sumner speaks, would have been greater than it was.

The New England town and the Southern plantation achieved order and industry and some degree of civilization along the Atlantic tidewater. But in the back country, on the frontier further west, the "English Tartars," as Burke called the large number of squatters who settled there, were passing through the same sort of chaotic conditions through which the older settlements were emerging. Indentured servants or redemptioners moved to the edge of free land, made a clearing, and "squatted" there. Unlike a Southern plantation or a New England church community, they did not migrate or settle as a social group larger than the family. They did not seek to transplant an established political institution or to impose their culture upon the Indians. Their trade was largely barter trade on a small and local scale. Methods of warfare reverted to Indian methods, even to scalping.[43] Property struggle for became sacred, and its theft was punished with great severity. In short, it was just another individualism on another frontier.

> It was an individualism, [says Mead,] which placed the soul over against his Maker, the pioneer over against society, and the economic man over against his market. . . . The American pioneer was spiritually stripped for the material conquest of a continent and the formation of a democratic community.[44]

But the tobacco and cotton frontiers were not democratic communities, for competition on them was not all; there was racial conflict to be adjusted, and there was need of some such institution as the plantation to effect the adjustment.

3

The Plantation in Virginia

Free Land and Plantation Settlement

When Tacitus, the Roman, went into the land of the Germans he was impressed with the abundance of unappropriated land—*superest ager*. It was the "free" land of the Roman frontier. Perhaps, in the existing state of German culture, the resources of the country were being fully utilized, but to Tacitus, viewing the country with the eyes of one accustomed to Roman techniques and methods, it was an area of "open" resources, capable of far greater yields to sustain a far greater population.

Likewise it is possible that the resources of Virginia, from the point of view of the aboriginal Indians, were being utilized about as completely as possible when the English came upon the scene. But to the newcomers with their background of economic training and experience the resources of the country were far from being closed. Like Tacitus in Germany, the English in Virginia were impressed with the great abundance of undeveloped virgin land, wherefore Queen Elizabeth named the country for herself.

> Sir Thomas Dale . . . declared that his admiration of Virginia increased as his opportunities for informing himself about its resources enlarged and that he believed it would be equivalent to all the best parts of Europe taken together if it were only brought under cultivation and divided among industrious people. Percy was equally emphatic in asserting that if the promoters of the Virginia enterprise would only extend the adventurers a hearty support, the new country would be as profitable to England in time as the Indies had long been to the King of Spain. Whitaker describes it as a place beautified by God with all the ornaments of nature, and enriched with his earthly treasures. "Heaven and Earth," exclaimed Captain Smith, "never agreed-better to frame a place for man's habitation." Williams apostrophized it as Virginia the fortunate, the incomparable, the garden of the world which, although covered with a natural grove, yet was of an aspect

so delightful and attractive that the most melancholy eye could not look upon it "without contentment, nor be contented without admiration." . . . "Where Nature is so amiable in its naked kind," asks the author of Nova Britannia, "what may we not expect from it in Virginia when it is assisted by human industry, and when both art and nature shall join to give the best content to men and all other creatures?"[1]

To the English it seemed impossible to allow the right of uncivilized Indians to monopolize resources that they did not and could not utilize. "We chanced in a lande," said John Smith, "where we found only an idle, improvident, scattered people, ignorant of the knowledge of gold, or silver or any commodities."[2] One hundred years later Beverly, after enumerating the natural produce of the country, said:

> This, and a great deal more, was the natural Production of that Country, which the Native Indians enjoyed, without the Curse of Industry, their Diversion alone, and not their Labour, supplying their Necessities. The Woman and Children indeed, were so far provident, as to lay up some of the Nuts and Fruits of the Earth, in their season for their farther Occasions: But none of the Toils of Husbandry were exercised by this happy People; except the bare planting of a little Corn, and Melons, which took only a few Days in the Summer, the rest being wholly spent in the Pursuit of their Pleasures.[3]

Except for the loss of soil fertility and the depletion of the original forests, the physical features of Virginia are today very much as the Indians knew them and as the English found them. The soil of the country which the planters settled upon was composed of alluvial sediments washed down from the geologically older mountain and Piedmont regions to the west by rivers and streams. As compared with the soils of the Piedmont they are generally more sandy, more subject to washing and leaching, and more level in topography. There are today in Tidewater Virginia a wide variation in soils, "perhaps fifteen kinds occurring, each sufficiently distinct to constitute almost a series." In the Piedmont region above the fall line of the rivers, the soils generally consist of the residium left upon the decay of the underlying rocks and varies from an inch to fifty feet or more in thickness.[4] The soils which were to become especially favorable for tobacco cultivation were those composed of a moist and fertile mould fortunately lying along the banks of the navigable rivers.

In 1607 the planters became acquainted with Powhatan, emperor of thirty petty rulers of tribes. According to Macleod, he ruled over an area of about five thousand square miles of Tidewater Virginia and had subjects numbering about

10,000.⁵ They lived by agriculture, hunting, and fishing. In common with all the aborigines of eastern North America, their chief crop was maize. Maize was supplemented by a kind of millet, melons, and sweet potatoes. Tobacco and gourds were also cultivated. An oil derived from hickory and walnuts served as a food. Agriculture was woman's work; the men were hunters and fishers. Deer and various small game were abundant, and the streams afforded a large number of fish.

The Indians of this area seem to have held both chattel and debtor slaves, but unlike slavery among the Indians of the Pacific Northwest this slavery was not hereditary. "Slaves were merely captive men, women, and children who were either waiting adoption or were not adopted."⁶ Of the Indians of Virginia Beverly states:

> They have also People of a Rank inferior to the Commons, a Sort of Servants among them. These are call'd black Boys, and are attendant upon the Gentry, to do their servile Offices, which, in their State of Nature, are not many. For they live barely up to the present Relief of their Necessities, and make all Things easy and comfortable to themselves, by the Indulgence of a kind Climate, without toiling and perplexing their minds for Riches, which other People often trouble themselves to provide for uncertain and ungrateful Heirs. In short, they seem, as possessing nothing, and yet enjoying all Things.⁷

This briefly describes the land and the people which the first planters of Virginia encountered when they landed in April, 1607. Let us now turn our attention to the particular organization—the Virginia Company—which sent them there and sought to direct their activities until 1624.

The Virginia Company of London was created by letters patent from James I in 1606. It was organized as a joint-stock company on the same plan as the East India Company. More than one hundred of its shareholders were also shareholders in the East India Company and the same businessmen were active in both. The charter members of the Company were, for the most part, merchants of London. After its organization two classes of members were distinguished: (1) "adventurers," who remained in England and subscribed money toward a capital stock, and (2) "planters," who agreed to go in person and were expected by their industry or trade to enlarge the stock and its profits. Labor and capital were, apparently, on the same footing, but the share of each in the results of the enterprise was at first not very clearly defined. When the company was reorganized in 1609, however, the capital stock was divided into equal shares of £12 10s. One share of the stock was given to an ordinary planter above ten years of age for his "adventure of the person." The plantation was, in other words, a transplantation of a portion of the Company. The planters were also entitled to support from the Company for five, later changed to seven, years. Under the governorship of Sir Thomas Smyth the Company in the first twelve years expended £80,000.[8] This gentleman was, incidentally, after 1606, governor of the Muscovy Company, governor of the East India Company, governor of the Bermuda Company, and a director of the Levant Company, in addition to being governor of the Virginia Company. No better proof of the commercial motives which governed early relations with Virginia could be had.

Nevertheless, in the face of actual conditions of settlement the Company

worked away from its original purpose of securing wares from the natives, and came to depend upon the development of the resources of the country, making extensive grants of patents for private plantations and establishing free trade, but turning to monopolies of staples in the later years.

The objects and motives of the Virginia merchants are difficult to discover and hence to compare with those of other trading associations, in that they change with the progress of the company. In fact, it was by these changes this company worked out the distinction which was hereafter to be made between companies for trade and associations for colonization. It aimed at "ways of enriching the colonies and providing returns so that the fleets come not home empty," and hence there was a similarity of purpose with the East India Company in the endeavor to discover a route to the South Sea to find mines, and to secure trade as well as tribute from the natives. But here the ways parted and the orders to expand labor in producing wines, pitch, tar, soap, ashes, steel, iron, pipe staves, hemp, silk grass, and in securing cod, sturgeon, and pearls distinctly mark the path which the Virginia Company was of necessity to follow. It was to secure its wealth by the development of the country. But it finally reached the position of the African Company, which became an organization for securing a staple—namely negroes. For the Virginia Company was to find its sole resource and hope in the productions from its lands and the importation of tobacco.[9]

Between developments in Virginia and changes in company policy there was considerable lag; changes in the company followed after some time changes in the plantation's requirements and adjustments. Let us see how these changes in the plantation itself came about by the natural pressure of circumstances. They followed from the fact that the planters unexpectedly found themselves in an area of great agricultural possibilities but with little opportunity for trade on a highly profitable scale. Circumstances forced them to utilize some of these resources to sustain themselves, although they had been promised support by the Company. Hence the changes in the character of the plantation and of the Company itself begin with the acquisition of the territory, not in order to establish political over-lordship over the natives as the East India Company did in India, but to utilize directly the land and its resources.

Since the prevailing English opinion was that the Indians had no real interest in the lands of Virginia but only a general residence there, like the wild beasts of the country, there were few scruples against forcible appropriation. The Virginia Company, under the terms of its charter, considered itself as holding complete title to the country since the Indians were ruled by a non-Christian prince. Nevertheless, so long as trade was their principal motivation there was little need of actually taking possession of the land for settlement. But the acquisition of land became more important as the colony came more and more to rely upon agriculture, and especially upon tobacco. In spite of the fact that the Virginia Company claimed title to the land, it was expedient to acquire actual use of it by peaceful means. Jefferson stated that most of Tidewater Virginia was acquired, not by conquest but by the process of lawful exchange.[10] Bruce states that "the larger part of the Peninsula, the seat of the earliest English settlements, was acquired at first by conquest, but right of possession was afterwards confirmed by treaty."[11] By forcible entry and fraudulent devices the advancing tide of settlers gradually took possession of an ever-enlarging area. About 1660 the settlers, now thoroughly in control, attempted through the Assembly to reduce the Indian holdings within a definite limit. This came partly as a result of complaints from some tribes that they had been deprived of their lands to such an extent that they were in a straitened condition. A system of reservations was organized by the Assembly and the natives forbidden to alienate the land reserved to them. After a war between the colony and the aborigines in 1676, however, the efforts to safeguard Indian rights diminished, and most of the reservations finally disappeared. The Indian population declined, and great tracts became available without any removal of the native owners. The white man's diseases fought on his side far out in front of his line of settlement.[12]

With the land in the actual possession of Company and Crown it became relatively easy for the individual settlers to acquire it. By the time the private settler came into existence, however, the nature of the plantation had undergone some fundamental changes. We will see what these were in order to understand why land was wanted for settlement at all.

It is certain that what the Virginia Company intended to establish in Virginia was a trading and exploring factory. How else can be explained the character of the original planters whom they sent—"gentlemen" and tradesmen and but few men, if any, versed in agriculture? They settled on a marshy island on the northern bank of the James River, a very unpromising site for agriculture but one affording protection for a factory. They came without wives, for they were traders, and traders of whatever race or nationality find it good as well as satisfying business to take women from among the group with whom they trade. John Rolfe, one of the planters, married Pocahontas, the daughter of Powhatan. The profits of the venture

> were to be spent upon the settlement, and the surplus was either to be divided or funded for seven years. During that period the settlers were to be maintained at the expense of the Company, while all the product of their labours was to be cast into the common stock. At the end of that time every shareholder was to receive a grant of land in proportion to his stock held.[13]

Wingfield was the first "president" of the resident council of Virginia, a title which followed the practice of the English factories in the Old World. Like these factories, also, the resident council was given the power annually to choose its president, to make and enforce ordinances for the government of the settlement, so far as these ordinances were in keeping with English law, to sit as a court of justice, to appoint minor officials, and to exercise the functions of local administration. There was to be a Cape Merchant, or treasurer, to receive the goods sent to the colony and to sell those sent home. He was also to administer the common store.[14] This organization was maintained in theory until the reforms of Governor Dale.

A comparison of the organization of the first settlement at Jamestown with that of trading factories maintained by the English in the Baltic, the Levant, and in India would reveal the extent of the similarity. Such a comparison would make understandable the so-called "communism" of early Jamestown. That the hard-headed business men who composed the London Company are sometimes credited by American historians with an attempt to establish in Virginia something so foreign to the nature of capitalistic business is difficult to account for. It undoubtedly results from a failure to take into account the fact that the English pattern of commercial contact with oversea peoples was the trading factory, and the merchants of the Virginia Company erred in nothing more than in conceiving of Virginia in the same terms.[15]

It is probable that the Indians of the region understood the motives of the settlers at Jamestown and quickly responded to them. Macleod says:

> Powhatan was apparently glad to have an English trading community settled in his territory. He profited in trade, to the disadvantage of neighboring enemy tribes whom he would not permit to trade with the English. He

always endeavored to keep the colony from making contacts with powerful neighboring Indian peoples, fearing that as a consequence he would
lose his monopoly of the English trade. . . .[16]

So long as trade was the dominant motive of the plantation, the Virginia
Company and its governors at Jamestown insisted upon friendly relations with
the Indians. When John Smith and Strackey, the secretary of the plantation, who
foresaw the inevitable agricultural future of Virginia, desired to conquer and
enslave the natives by force of arms, the Company persisted in its policy. The
Company desired the greatest amount of trade with the least amount of expense.[17] Indian and English attitudes toward each other began to change with the
change in motives on the part of the planters from trade to agriculture.

This transition began, by necessity, during the first year of the plantation.
Until the fall of the year they fared well, but disease and falling provisions brought
on a "starving time." Wingfield, the first president, remembering the obligation
of the Company to maintain the planters for five years, counselled husbanding
the small store of provisions until help arrived from England. The starving settlers
arose against him and elected John Smith, whose resourcefulness had attracted
attention, president. Smith's forceful command of the colony and his decree that
those who would not work should not eat is well known from our school histories. For this factory of Jamestown was to be maintained in a wilderness under
conditions far different from the factories in the coast cities of the Levant and
India. To cultivate the soil and not to trade, as John Smith demanded, was to turn
the factory toward industry and away from commerce. To enforce his orders
Smith resorted to a measure not unknown to modern plantation and mining
companies in the United States—control through the commissary or company
store. A company of traders may be organized under a "president," but an agricultural plantation requires an overseer with full authority. Hence, in 1609, government by council was abandoned and a governor appointed with the authority
of military law. He was the mild Lord Delaware who returned to England in 1611
but continued to rule through deputy governors, or overseers. From 1611 to 1616
this office was filled by Governor Dale, who found it necessary to reestablish
Smith's methods and did so with great severity.[18] When Dale left the colony in
1616 agriculture was thoroughly established as its main industry.

Two weeks after settlement the colonists planted a small clearing in wheat
which they had brought with them from England, but it is evident they were not
prepared to depend on wheat. Previous to 1609, when the provisions supplied by
the Company ran low, they were absolutely dependent upon the Indians, who
supplied them with corn. In this year John Smith introduced the culture of maize
among the colonists. The art was learned from the Indians and forty acres

planted. Agriculture apparently declined during the succeeding administration of Lord Delaware, whose chief dependence for food was placed upon the Company and upon trade with the Indians. Maize culture before the arrival of Governor Dale was almost abandoned. Dale modified the "communistic" organization of the factory to meet conditions by assigning a separate garden to each man and laying off a common garden to be devoted to the cultivation of hemp and flax. Dale's agricultural interests led him to found the town of Henrico in 1611 and to extend the area under immediate control of the English. He seized the land of the Appomattox Indians and divided it into Hundreds. On the frontier he established a line of individual homestead settlements to protect the Hundreds against the Indians. This new area was primarily needed as a range for the cattle which Dale brought with him from England as well as for additional maize cultivation.

In 1612 the first tobacco was cultivated by the planters. Like [the acquisition of] corn, the acquisition of tobacco from the Indians was very uncertain and irregular, and the experiments of John Rolfe in attempting its cultivation were probably undertaken with a view to providing a more regular supply. The success of his experiments, according to Bruce, would "have led to the exclusive cultivation [of tobacco] by the colonists, but for the fact that Sir Thomas Dale was able to govern their action."[19] Dale continued to insist upon a subsistence agriculture. But with the trend toward private ownership in land the cultivation of the marketable staple which had been discovered grew rapidly.

The "private gardens," or farms, assigned by Dale to a number of the planters were three acre tracts to be held under lease from the Company. In addition to [paying] the rent, each tenant was required to work for the commonwealth one month in the year. No tobacco might be planted until the tenant had put in two acres of grain. The rest of the colonists were required to work in the common garden to provide for the common store for eleven months of the year. This step in 1613 was taken in connection with the founding of Bermuda and marks the first tendency in the colony toward private ownership. In 1616 certain planters were given permission to hire labor from the colony. After 1617, associations of planters began to be formed which acquired certain corporate rights to land. General private ownership was established by Governor Yeardley in 1619, who assigned, in lieu of dividends from the Company to the planters, fifty and one hundred acre tracts to those entitled to receive them.

When it became apparent that no money dividends could be expected from the Virginia Company, large shareholders in England, such as Lord Delaware, asked for and received grants to land instead. They formed independent associations for the purpose of planting the lands to which they were entitled. Land patents to such joint-stock associations were for "particular plantations." In April, 1625, forty-four patents had been granted to persons or associations who undertook to transport at least one hundred men each to their plantations. After 1619 most of these patents were to single individuals. The largest of these "particular" plantations settled was Smyth's Hundred, an area of about 200,000 acres and with 310 persons, in May, 1620. These joint-stock plantations, however, were not very successful and gave way to private planters, many of them able to add to their original grants of fifty or one hundred acres by purchase, headright, or fraud. The Company reserved to itself 12,000 acres, a part of which was cultivated directly for the Company by tenants sent out from England. It also became a practice of the Company to maintain their officials in Virginia without any need of their reliance on the resources or the Company in England by special land grants, together with tenants to cultivate them. With the establishment of every new office went a certain number of acres to maintain it.

Even before the dissolution of the Company in 1624 every emigrant from England who transported himself at his own expense to the colony was declared entitled to fifty acres of land. In addition every resident of Virginia who imported an immigrant was entitled to fifty acres known as a "headright." Headrights were saleable and subject to fraudulent manipulation in the interest of the influential planters. With the development of indentured servitude, planters became entitled, not only to fifty acres for each servant imported, but to the labor of the servant for a period of years. Headright and indentured servitude together were efficient measures in settling the land and bringing it into cultivation. As early as 1623 there were many private plantations of as much as 5,000 acres, and by 1635 there were planters whose landed possessions amounted to 10,000 acres each, but the average land patent up to 1650 was about 450 acres. After 1650 the yearly average of land patented by individuals began to grow larger, and private plantations of from 20,000 to 50,000 acres were acquired. When slavery was substituted for indentured servitude, the principle of "headright" became obsolete, and a policy of making grants of land by the Virginia Council took its place. The grants of the seventeenth century were, for the most part, not very large, but the large grant gradually became more frequent.

> Between 1695 and 1700 there were seven patents for from five thousand acres to ten thousand acres, and in 1726 alone grants of from two thousand acres to forty thousand acres became common. And most of these grants were to men who already held lands. By the middle of the century estates of over a hundred thousand acres were not unknown. Most planters shared William Fitzhugh's willingness to purchase whatever available lands they thought "convenient" and a few were not averse to the use of means somewhat questionable to add to their already extensive holdings. The story of the rise of the estates of the Byrds, Joneses, Fitzhughs, Masons,

Washington, Carters, Lees, Epeses, Beverlys, Allens, etc., is one of continued additions by government grant, headrights, or purchase from individuals. The motives were of course various, but one ever in the background
was the want of timber and elbow room for future agricultural expansion.
On most of these estates we have the record of lands "somewhat cultivated"
being left behind as planting was pushed into new areas.[20]

A high degree of isolation naturally followed from the existence of such large
agricultural estates. The Reverend Peter Fontaine wrote in 1754 that about a
thousand acres "to keep troublesome neighbors at a distance" and a few slaves to
make corn and tobacco and a few necessities were sufficient for his wants. A character in Kennedy's *Swallow Barn* who opposed a good roads bill recently introduced in the General Assembly declared that "the home material of Virginia was
never so good as when her roads were at their worst." But the Virginia plantation
was not completely isolated, for contact with the market and some means of
transportation for its staple were important considerations.

In the seventeenth century . . . the area included in the patents was confined principally to the lands which were situated immediately on the navigable streams. The number of these streams were extraordinary. Beginning
with the Powhatan, York, Rappahannock, and Potomac, there were, at
comparatively short intervals, rivers, creeks, or estuaries deep enough to
float the largest ships employed in the carrying trade between England and
Virginia. At that early period every planter owned a wharf; indeed the
strongest reason after fertility of the soil which influenced him in selecting
a tract of land was that it fronted on a water highway.[21]

As the Virginia plantation turned its attention from trade to agriculture, there
was a corresponding change in attitude of the planters, not as factors, but as laborers. John Smith, as we have seen, introduced this new status for the planters, but
until the governorship of Dale it was regarded as temporary. Legally the planter
stood on an equal footing with all other members and stockholders of the Virginia
Company, but actually he became little more than a servant in the general employ
of the Company, forbidden to leave the plantation if the governor required him
to stay and with no right of residence if the governor wanted him to leave. For a
few years the planters worked as hirelings on the Company's lands, "having most
of them served the colony six or seven years in the general slavery," as one of them
wrote.[22] When Dale was sent out as overseer of the plantation in 1611, it was with
instructions to "rationalize" the business, to use a modern term. A gang system of
labor was adopted under which the planters were divided into squads, each under
the command of a captain. Under Dale the fiction of stockholder equality in

the Virginia Company came to an end and the principle of coercion definitely recognized.

When planters themselves were established upon their private plantations they in turn became employers and sought labor from whatever source available. The extent to which Indians were used as servants and slaves is not very well known, but it is probably that they began to be so used after the Indian massacre of 1622. But they made very unsatisfactory laborers; the universal judgment was that they could not endure the yoke of regular labor and were natively too proud and fond of liberty to make good servants or slaves.[23] It soon became evident that the tobacco plantations of the planters would have to be cultivated by labor imported from elsewhere if they were to be cultivated at all. For tobacco was absorbing all the thoughts of the Virginia planters in those days.

Agricultural Specialization: Tobacco

Tobacco possesses an adaptability to a wide range of geographic and climatic conditions, both tropical and temperate. It is also highly sensitive to variations in soil and climate, perhaps more than any other major staple crop. This accounts for the wide distribution of cultivation throughout the world, with each area having a certain degree of immunity from the competition of other areas. Virginia, Kentucky, Connecticut, Cuban, Sumatran, and Turkish tobaccos are all distinctive types which depend finally upon the soil and climate of the areas in which they are grown.[24] The commercial tobacco sold on the market today usually represents various mixtures of these types, which accounts for the fact, unusual in agriculture, of producing areas importing tobacco from other producing areas.

> As tobacco culture was carried from the first settlement at Jamestown into new territory, it was seen that the changes in soil and climate resulted in important differences in the character of the tobacco produced. It gradually became more and more apparent also that these differences in the properties of the tobacco leaf, due to soil and climatic influences, greatly affected its adaptability for use in different forms; the product of one section, for example, was especially suitable for making smoking or chewing tobacco, but perhaps did not produce so acceptable a cigar as that of another section. It was learned, moreover, that desirable characteristics of the tobacco leaf resulting from local soil and climatic influences could be further accentuated by modifying the methods of growing and curing. Thus, through a process of gradual evolution, tobacco culture has become highly specialized, each producing district furnishing a distinctive type especially adapted for certain uses, based ultimately on the tastes and preferences of the consumer. It is the accumulated experience of three centuries

of tobacco culture that each of these types can be produced only under certain conditions of soil and climate, by using certain varieties of seed, and by employing special methods in growing and handling the crop.[25]

There was a sort of vitalism about the tobacco plantation in Virginia, as tobacco from that colony came to compare favorably with the Spanish tobaccos of the West Indies.[26] It possessed a certain industrial potential which sprang into life when the proper connection with a European market was made. "For no other Virginia product," says Bruce, "was there opportunity for a sale that would enlarge as the amount exported increased." Tobacco, like most luxuries and especially stimulants, increased its marginal utility with successive increments of supply.

Despite all obstacles this particular territorial division of labor in the world community insisted upon being born. In the face of opposition, and probably because of it, the demand for tobacco in England grew steadily. Production in Virginia went rapidly forward despite the wishes of the Virginia Company which had sought in the New World the "solid commodities" of commerce. Kings James I and Charles II constantly opposed it. It was to the dishonor and shame of the people of Virginia, Charles declared, that the plantation was built upon smoke alone; "the king is careful to encourage and support the plantation and he has long expected some better fruit than tobacco and smoke to be returned from thence."[27]

Nevertheless as tobacco imports grew in England the duties helped to bring needed money into the Royal treasury, and it became politic to aid the English tobacco plantations. In 1625 the House of Commons voted to cut off the importation of competing tobaccos from Spanish colonies. The preferential tariff in favor of Virginia remained during the colonial period as a great support of the Virginia planters. In addition, tobacco-growing in both England and Ireland was prohibited in 1652.

Four years after John Rolfe's experiments in the cultivation of tobacco in 1612, it had already become one of the staple crops of the colony. Governor Dale in 1614 had discouraged the industry by insisting upon the raising of food-stuffs, but by 1617 the *George* set sail for England laden with 20,000 pounds of Virginia leaf—"the first of the vast fleet of tobacco ships which for centuries were to pass through the capes of the Chesapeake bound for Europe."[28] This first consignment sold in England for 5s. 3d a pound, a price which promised for the first time a substantial return to the Virginia Company.

No wonder the leaders of the London Company were pleased, believing that in the Indian weed they had discovered a veritable gold mine! No wonder the settlers deserted their palisades and their villages to seek out the richest soil and the spots best suited for tobacco culture! The man who could produce 200 pounds of the plant, after all freight charges had been met, could clear some £30 or £35, a very tidy sum indeed for those days. It was the discovery that Virginia could produce tobacco of excellent quality that accounts for the heavy migration in the years from 1618 to 1623. In fact, so rich were the returns that certain persons came to the colony, not with the intention of making it their permanent residence, but of enriching themselves "by a cropp of Tobacco" and then returning to England to enjoy the proceeds.[29]

The growth of population and the continuation of the tobacco excitement led to a proportionate increase in exports following 1618. From the James River Valley cultivation spread to the York, Rappahannock, and Potomac River Valleys and then along the Chesapeake clearings in Maryland. The extension was also southward into the Albemarle and Pamlico sections of North Carolina. The rise and decline of tobacco production and the wide fluctuations in prices received by the planters between 1619 and 1775 are shown in the accompanying table. The story behind this table is a more detailed and complicated one than can be reviewed here but some phases of it may be touched upon.

So long as the settlement at Jamestown was regarded as a factory of the Virginia Company it was, of course, the function of the Company itself to market whatever goods were acquired. With the relative change from trading to agricultural motives and the right to export goods independently of the Company, which the planters acquired in 1618, there was a fundamental change in the economic base and the transportation system of the colony. This change was immediately reflected in the social and political organization leading directly, as we have seen, to the plantation large estate and various forms of forced labor. The nexus between the industrial organization of the colony and its social and political superstructure was in Virginia, as everywhere in the world community, the marketing system. The marketing system was the first to respond and readjust to the change in the life conditions of the colony, and constituted an index of further social and political changes.

Table I

A Recapitulation of the Quantity and the Price of Tobacco Exported from Virginia from 1619 to 1775

Tenth Census. III, 224.

Year	Crop, no. lbs.	Prices per lb.	Per Cwt.
1619	20,000	3s	
1620	40,000	8d to 2s	
1621	55,000		
1622	60,000		
1628	500,000	3s to 4s	

Table I *continued*

Year	Crop, no. lbs.	Prices per lb.	Per Cwt.
1632		6d	
1633		9d	
1639	1,500,000	3d	
1640	1,300,000	12d	
1641	1,300,000	20d	
1661			12s
1662			10s
1664		(a)	
1666	(b)		
1667	(c)		
1682		10s	
1687	(d)		
1688	18,157,000		
1704	18,295,000	2d	
1729		2d	
1731	(e)		
1739			12s 6d
1745	38,232,900		14s
1746	36,217,800		(f)
1747	37,623,600		(f)
1748	42,104,700		(f)
1749	43,188,300		(f)
1750	43,710,300		(f)
1751	42,032,700		(f)
1752	42,542,000		(f)
1753	53,826,300		(f)
1754	45,722,700		(f)
1755	42,918,300	(g)	10s
1756	25,606,800		
1757	(h)		
1758	22,050,000		50s
1759			16s 8d
1760– 1775	(i)		18s to 25s

(a) 3d to 3½d per pound, the price in London.
(b) No crop made. Planting prohibited.
(c) Two-thirds of the crop destroyed by a storm.
(d) Poor crop of bad quality.
(e) Exports of Virginia and Maryland reported as 36,000,000 pounds.
(f) Price current, at which estimated according to contemporaneous records, 1746–1764, was 2d per pound.

(g) Price in London, 11d to 12½d.
(h) A short crop made.
(i) Average about 55,000 pounds annually. Price about 22s 6d per cwt.

The vessels and factors of the Virginia Company initiated a transportation and marketing system between Virginia and the Old World. In 1618, as stated, individual planters were permitted to export and [in] 1619 English merchants, not members of the Company, were allowed to engage in trade with Virginia. In 1624 Dutch vessels began to haunt the shores of Virginia and to transport tobacco at lower rates and to find a market for Virginia tobacco in the Netherlands. Factories and storehouses had been set up in Virginia at Middleburg and Flushing in 1621 to handle the growing tobacco trade. In 1624, when the Virginia Company was dissolved, a royal monopoly designed to restrict the quality and quantity of Virginia tobacco and regulate its trade was established. [King] James attempted to fix the price at 3 pence per pound.

In spite of all these attempted restrictions the Virginia tobacco industry grew enormously. "At Christmas 1648, . . . there were trading in the colony ten ships from London, two from Bristol, seven from New England, and twelve from Holland."[30] But overproduction, that nightmare of all plantation agriculture, was upon the planters, and prices fell below the cost of production. The competition of other producing areas was beginning to be felt. The following account of the tobacco industry of the period reads very much like present-day troubles of cotton planters:

Over-production seems to have been a constant source of trouble for the Virginia planters. To check this, as well as to prevent the fall in price, numerous acts were passed by the Assembly. Prices fell from three shillings per pound in 1620, to three pence per pound in 1640. During this period, not only did the Assembly fix the price of tobacco in terms of English money, but it also fixed the price of other commodities in terms of tobacco. Finding that the fixing of prices failed to remedy matters, the government tried other means of State regulation. It attempted to limit the supply by fixing the maximum number of pounds each planter could produce per cultivator employed. Another method resorted to in order to increase prices, was the destruction, by government inspectors of the poor grades of leaf. Finally, the condition of the market was so bad, and the debts could be legally cancelled upon payment of forty per cent (forty cents on the dollar) in terms of tobacco, the price of which was already fixed by law.

Having secured only temporary relief by enactments directly regulating tobacco, indirect means were resorted to. Colonial authorities, as well as

Parliament, tried to induce the colonists to substitute other crops for tobacco. Flax, hemp, cotton and silk were tried but these yielded an inadequate return. Even shipbuilding and trading were resorted to, but these
also proved poor substitutes. The trouble with all these artificial regulations was, as the colonists themselves saw, that Maryland was able to increase her output when Virginia attempted to curtail her own. And when
selling prices were fixed too high, English merchants would buy of Maryland. Besides, Spanish and Dutch traders were bringing tobacco from
the West Indies to the continent. Virginia planters tried to get Maryland
planters to agree to some plan whereby prices could be controlled. It was
suggested that in years following heavy crops all production should cease
in both colonies. Owing to mutual suspicion this plan, tried in 1666–
1667, fell through. The poor farmers of Maryland, said Lord Baltimore,
could not stand a year's cessation of crops, especially since their farms were
mortgaged. It should be added that, had the plan succeeded, Lord Baltimore would have suffered a loss in his revenues which came from tobacco
export duties and a tobacco poll tax.[31]

In 1651, 1660, and 1663, the English navigation acts regulating trade were
passed, excluding the ships of foreign countries from carrying the commerce
between England and her colonies. They were sufficiently enforced to raise the
charges of tobacco transportation and bring strong protests from the planters. In
1657 freight rates rose from £4 sterling to a ton to £9 and even £14. Nevertheless,
the exports from Virginia continued to increase.

In 1669 England sought to organize and systematize her policy with reference
to her trade with the plantations. A Board of Trade and Plantations was appointed
to enforce all previous regulations and to make new ones as required. The Board
began by adding to the duty levied on tobacco imported into England, bringing
the total to six-pence per pound. In time these duties came to constitute about
seven-eighths of the charges that the planter had to pay. In addition the factor's
commission and freight rates had to be paid before the planter received any return
from his crop. Colonial duties, laws relating to inspection, weighing, stamping,
and exporting imposed further restrictions on the planter's economic liberties,
but, in spite of all, the momentum of tobacco production could not be stopped.
The exports increased each year in the hope that unknown markets might be
found. Exports increased to the very beginning of the Revolution,[32] with two or
three hundred ships annually engaged in the trade.

In the meantime the factory had undergone an interesting transformation. It
had come to trade with the Indians, but it remained to trade with the planters,
although it had no connection with any one particular English Company. Factors

representing Glasgow merchants, for the most part, were carrying on the commerce.[33] To these merchants the planters were usually in debt. After paying marketing charges, the planter received, according to MacPherson, less than 25 per cent of the selling price of his tobacco.[34] Out of this slaves must be purchased and supported, fresh lands for tobacco cultivation acquired, and other production expenses paid. It is evident that the plantation was paying diminishing returns on the English capital for which the planter, not the merchant from whom it was borrowed, suffered the losses. A frantic search for a staple to take the place of tobacco was made during the eighteenth century, and although lumber, wheat, corn, hemp and other products were exported in large quantities, the main reliance continued to be on tobacco. But the peak of prosperity based upon the trade in that staple had passed, and the Virginia plantation seemed on the decline.

Diminishing returns in the agricultural industry of a particular area may be an index of changing spatial distance when measured in terms of cost.[35] In the parts of the community where commercial agriculture is, or may be, pursued, the various crops are competitors, or potential competitors, within natural limitations of soil and climate, for space. In the competition certain advantages seem to rest with certain classes of agriculture, depending upon cost of transportation to market.[36] It was not the *Nicotiana rustica* which the planters found the Indians growing, but a variety of *Nicotiana tabacum* obtained from South America and the West Indies that became commercially profitable.[37] But by 1800 changed market relations had so depressed the prices of tobacco and exploitive methods of cultivating tobacco had so exhausted the fertility of the soil that the planters were searching for another staple and seemed about to find it in wheat.

In the modern city, people compete for space and for standing-room and their resulting distribution tends to take a characteristic pattern around the city center.[38] In the agrarian parts of the community the various types of agriculture, as Thünen has shown, compete for space and the resulting distribution around the central market place likewise tends to follow a characteristic pattern. In finding their economic positions, however, these various types of agriculture select and distribute population, their caretakers, with them. White indentured servants, Negro slaves, and the planter class of Virginia may be considered from this point of view.

4

~

Plantation Management and Imported Labor in Virginia

The Tide of White Labor

The shift in ecological position which made England the center of an evolving world community brought her into relation with Virginia, and resulted in attempts to follow up that relation in terms of past experience with oversea peoples through trading factories. When the trading plantation or factory became, through force of circumstances, an industrial plantation producing an agricultural commodity, the nature of the relationship changed and became at the same time a more integrated one. One aspect of the new relationship was immigration rather than transplantation. The change was evidence of a further shift in the ecological positions of England and Virginia. Immigration was the consequence of the new spacial and commercial relation.

The year 1619 is a memorable date in Virginia's history. In that year "twenty negars," to use the language of John Rolfe's journal, were landed from a Dutch vessel at Jamestown. In 1619 a representative group of Virginians met for the first tine in the capacity of a General Assembly with power to legislate for the colony. In 1619 there arrived at Jamestown a shipload of English women who became wives of those planters paying the Virginia Company in tobacco for their passage. They are witness to the fact that the planters were becoming colonists instead of factors. In 1619 one hundred pauper boys and girls from the London streets were sent to Virginia and bound out to the planters for a term of years.

All of these events of 1619 bore a significant relation to each other and to the future. The building up of a resident planter class, whose business of producing tobacco for the English market was an increasingly profitable one, gave the plantation owners the new role of labor immigration agents. No longer were those who came to Virginia "planters," in the original pioneering sense of that term, but "immigrants" who were received into a social and economic system already established. "Immigration, like everything else in the colony," says Wertenbaker, "was shaped by the needs of tobacco. For its successful production the plant does not require skilled labor of intensive cultivation. . . . But it does require the service of many hands."[1] Indentured servitude, the form which this English immigration took, was a system of labor "around which the economic life of Virginia centered for a full century."[2] Throughout the seventeenth century the annual arrivals averaged between 1500 and 2000 persons. "All in all, considerably more than 100,000 persons migrated to the colony in the years that elapsed between the first settlement at Jamestown and the end of the century."[3] In 1619 secretary John Pory wrote that "one man by the means of six servants hath cleared at one crop [of tobacco] a thousand pounds English . . . our principal wealth consisteth of servants."

The continuation of the settlement of America by means of indentured servants was a result, like the original plantations earlier, of the unsettlement of Europe. Jernegan sums it up for us in the following excerpt:

> In the sixteenth century English agriculture was giving way to sheep-raising, so that a few herders often took the place of many farm laborers. As a result the unemployed, the poor, and the criminal classes increased rapidly. Justices, who were land landowners, had the power to fix the maximum wages of farm laborers. Sometimes they made them very low, hardly a shilling a day; for the lower the wage the greater the profits of the tenant farmer, and therefore the greater his ability to pay higher rents demanded by the landowner. Thus, whole wages remained practically stationary, wheat multiplied in price nearly four times in the period 1500–1600. In other words, a man worked forty weeks in 1600 for as much food as he received in 1500 by working ten weeks. To prevent scarcity of farm laborers, the statute of apprentices (1565) forbade any one below the rank of yeoman to withdraw from agricultural pursuits to be apprenticed to a trade. Moreover, the poor laws passed in this period compelled each parish to support its poor, and provided penalties for vagrancy. Thus the farm laborer had no chance to better himself. Conditions were almost beyond description and in dear years people perished from famine. Sheffield in 1615, with a population of 2,207, had 725 relying on charity, 37.8 per cent of the population.[4]

The apprenticeship laws sought to deal with pauper children by apprenticing them to masters for a period of years, usually seven, to be taught a trade. In addition to the term of service such details as wages and hours of labor were minutely prescribed. The one hundred pauper boys and girls sent to Virginia by the Virginia Company in 1619 were sent under a special application of Elizabeth's Statute of Apprentices of 1563, which put a premium upon agricultural apprenticeship. In this way the folkway of apprenticeship was gradually modified to create the pattern of indentured servitude. The chief difference, aside from the fact of overseas transportation, came to be a matter of age groups: apprentices were usually minors, whereas indentured servants were usually adults.[5] Some of the instruments of indenture read very much like a serf taking an oath of fealty to his lord. The following is typical:

> This Indenture made the 6th day of June in the year of our Lord Christ 1659, witnesseth, that Bartholomew Clarke ye Son of John Clarke of the City of Canterbury, Sadler, of his own liking and with ye consent of Francis Plumer of ye City of Canterbury, Brewer, hath put himself apprentice unto Edward Rowzie of Virginia, planter, as an apprentice with him to dwell from ye day of the date above mentioned unto ye full term of four years from thence next ensuing fully to be complete and ended, all which said term the said Bartholomew Clarke well and faithfully the said Edward Rowzie as his master shall serve, his secrets keep, his commands most just and lawful he shall observe, and fornication he shall not commit, not contract matrimony with any woman during the said term, he shall not do hurt unto his master, nor consent to ye doing of any, but to his power shall hinder and prevent ye doing of any; at cards, dice or any unlawful games he shall not play; he shall not waste the goods of his said master nor lend them to anybody without his master's consent, he shall not absent himself from his said master's service day or night, but as a true and faithful servant, shall demean himself, and the said Edward Rowzie on ye mystery, art, and occupation of a planter which now . . . the best manner he can, the said Bartholomew shall teach or cause to be taught, and also during said term shall find and allow his apprentice competent meat, drink, apparel, washing, lodging with all things fitting for his degree and in the end thereof, fifty acres of land to be laid out for him, and all other things which according to the custom of the country is or ought to be done.[6]

Indentured servitude in Virginia was both voluntary and involuntary. Voluntary servitude was based upon free contract with the Virginia Company "or with private persons for definite terms of service, in consideration of the servant's transportation and maintenance during servitude." Involuntary servitude followed when "legal authority condemned a person to a term of servitude necessary for reformation or prevention from an idle course of life, or as in a reprieve from other punishment for misdemeanors already committed."[7] In the latter case the contract of indenture was between the authority imposing the sentence and the planter who guaranteed the transportation of the servant. Servants of this latter class were paupers, vagrants, and political criminals. But with the additional exception of a number of servants who entered servitude by reason of being kidnapped or "spirited," the great body of servants seems to have voluntarily entered into the contract in order to get to Virginia. The contract was usually made with shipmasters who traded in Virginia and transported tobacco on the return voyage. On arrival in Virginia the shipmaster or his agent advertised the sale of servants and assigned his contracts to the purchasers. The following advertisement which appeared in the Virginia *Gazette* for March 28, 1771, is typical.

> Just arrived at Leedstown, the Ship Justitia, with about one Hundred Healthy Servants. Men, Women and Boys, among whom are many Tradesmen—viz. Blacksmiths, Shoemakers, Tailors, House Carpenters and Joiners, a Cooper, a Bricklayer and Plasterer, a Painter, a Watchmaker and Glazier, several Silversmiths, Weavers, a Jeweler, and many others. The Sale will Commence on Tuesday, the 2d of April, at Leeds Town on Rappahannock River. A Reasonable Credit will be allowed, giving Bond with Approved Security to Thomas Hodge.[8]

It is impossible to date the beginning of indentured servitude in Virginia. Some of the first gentlemen "planters" took servants with then on stipulated wages without contract of indenture. After 1616, resident planters began to import servants but whether or not these were indentured is uncertain. But it is certain that formal indentures were in use in the year 1619, for in that year, as we have already mentioned, one hundred pauper boys and girls were apprenticed to planters, and the first General Assembly held in the same year gave the practice legal sanction and also provided for the strict recording and performance of all contracts between both parties.

In the subsequent history of indentured servitude in Virginia, Ballagh distinguishes three general periods. The first was the period from 1619 to 1642, which was characterized by the fixation in custom of practices originating during the Company's government. In this period contracts might or might not be written. Where a verbal contract alone existed it was knows as "Servitude according to the Custom" and enforced by the courts in terms of customary practices. In 1643 the Assembly fixed a definite term for all servants brought in without indentures. This was known as "Servitude by act of Assembly" and marks the beginning of the second period, which continued until 1726. In the second period the Assembly by a series of statutes attempted to reduce servitude to legal uniformity. It was during this period that servitude in the colony reached its peak of importance. The third period, from 1726 to 1788, marked the decline of white indentured servitude and the rise of Negro slavery.[9]

For our purposes we are interested in tracing through these three periods the development of features about the system of indentured servitude which are of greatest significance for the social and political control of industrial labor, both white and Negro. Three developments seem especially significant: the assignment of contracts, the lengthening of servitude, and corporal punishment.

The assignment of contracts by deed or will—the legal forerunner of the slave trade, developed as a customary practice and was recognized by the courts before it was established in legislation.

> One of the earliest and most important customs was the right assumed by the master to assign his servant's contract whether he gave him consent or not. This originated in the practice with the Company of disposing of apprentices and servants to planters on their agreeing to reimburse the Company for the expenses of the servant's transportation, and in the custom with officers of the government of renting their tenants and apprentices to planters in order to insure an easier or more certain support. The depressed condition of the colony following the Indian massacre of 1622 made the sale of servants a very common practice among both officers and planters. In 1623 George Sandys, the treasurer of Virginia, was forced to sell the only remaining eleven servants of the Company for mere lack of provisions to support them, and a planter sold the seven men on his plantation for a hundred and fifty pounds of tobacco. The practice was loudly condemned in England and bitterly resented on the part of servants, but the planters found their justification in the exigencies of the occasion, and their legal right to make the sale seems never to have been actually called into question. Assignments of contracts for the whole or the unexpired portion of the servant's term became from this time forward very common. As a result the idea of the contract and of the legal personality of the servant was gradually lost sight of in the disposition to regard him as a chattel and a part of the personal estate of his master, which might be treated and disposed of very much in the same way as the rest of the estate. He became thus rated in inventories of estates, and was disposed of both by will and by deed along with the rest of the property.[10]

About the beginning of the eighteenth century a practical limitation was put upon the absolute right of assignment of contract by incapacitating "Jews, Moors and Mohometans" and all others who could not give "christine care and usage" (i.e., free Negroes, mulattoes and Indians), from owning Christian servants, but these might own non-Christian servants. In 1785, when the system of indentured servitude was practically at an end, the servant was given the right of assent to the assignment of his contract, to be attested in writing by a justice of the peace.

Offenses for which the servant might be punished included offenses against the community and offenses against the master. For the first, punishment was a penalty; and for the second, it partook of the nature of damages to the master for his loss. Since the master had virtually paid the servant his wages in advance when he paid the cost of his transportation to Virginia, dismissal for offenses was doubly out of the question; it would have equaled throwing the rabbit in the briar patch. Under these circumstances punishment inevitably took the form of lengthening the period of servitude, and corporal punishment.

What sort of offenses most frequently required punishment? They were, in general, the sort which violated the master's right to the full time and services of the servant under the terms of the indenture. Idleness was a frequent complaint. The right of free marriage would, especially in the case of female servants, work to the disadvantage of the master and consequently was forbidden. In 1619 the first General Assembly forbade a female servant to marry without the consent of her master, but the right of a male servant to marry was unrestricted until 1643, when a law was passed restricting it. In 1662 further restrictions on servant marriages were made. Running away was the most frequent offense. Even when the servant was recovered, the loss of time and expense was serious to the master. Offenses that partook more directly of infringements against the peace and order of the community included such offenses as robbery, rebellion, and fornication.

Punishment by means of extension of time had its beginnings in Virginia in 1619, when the first General Assembly ordered servitude for wages as a penalty for "idlers and renegades." This was to be service to the colony in public works and meant, of course, service in addition to the term of his contract. "In this we have the germ of additions of time, a practice which later became the occasion of a very serious abuse of the servant's rights by the addition of terms altogether incommensurate with the offenses for which they were imposed."[11] Since the servant usually did not possess the means of discharging a fine [as did] the freeman with property, other means of punishment were all the more necessary.

The law of 1643 provided that "What man servant soever hath since January 1640 or hereafter shall secretly marry with any maid or woman servant without the consent of her master he or they so offending shall in the first place serve out his or their tyme or tymes with his or their masters—and after serve his master one complete year more for such offense committed," and the maid "shall for her offense double the time of service with her master or mistress."[12] Other offenses connected with relations between the sexes among the servants brought added time. "When the offender was a free man he had to pay double the value of the maid's service to her master and a fine of 500 lbs. of tobacco to the parish. For the offense of fornication with a maid-servant the guilty man was required to give her master a year's service for the loss of her time, or, if a freeman, he might make a money satisfaction."[13]

In the case of runaways all expense incurred in the course of capture and return to the master, together with damages for the master's loss of time, was ultimately charged against the servant. Hence additions of time were usually greatest for this offense. "Additions thus frequently amounted to as much as four or five years, or even seven in some cases, and were often more than the original terms of servitude."[14]

It is significant that, aside from punishment for offenses at all, there were a number of instances of servants suing for their freedom who were held or sold for periods longer than their indentures called for. With labor so greatly in demand the offenses were not all on the side of the servant. Such unlawful holding of servants beyond their time of service was easier and more frequent in the case of Negro servants but was not unknown in the case of white servants.

Corporal punishment was provided for in a law of 1619 which read, "if a servant willfully neglect his master's commands he shall suffer bodily punishment." Until 1662 the right of corporal punishment was in the hands of the Assembly and the courts but was probably also exercised by masters without the legal rights.[15]

The extension of this important power beyond the administration of the courts was largely a result of the necessity of providing some severe correction in the case of runaways. The servant had generally no means wherewith to remit a fine, and so in penal offenses, where free persons were fined, we have seen that the servant was whipped, unless his master discharged the fine. In many cases also it was a general punishment both under the laws of England and under those of the colony, so that a law of 1662 provided for the erecting of a whipping-post in every county; but even before this time the master had assumed the right of administering corporal punishment to his servants. In this year it became a right recognized by law, but when a master received an addition of time for his servant's offense it remained doubtful whether corporal punishment could also be administered. This question was settled by the Assembly in 1668. It was declared that "moderate corporal punishment" might be given to runaways either by the master or by a magistrate, and that it should "not deprive the master of the satisfaction allowed by law, the one being as necessary to reclayme them from perishing in that idle course as the other is just to repaire the damages sustained by the master." The power thus given was doubtless abused, for in 1705 an act was passed restraining masters from giving "immoderate correction," and requiring an order from a justice of the peace for the whipping of "a Christian white servant naked," under penalty of a forfeit of forty shillings to the party injured. The act is significant as showing also the master's right to employ corporal punishment as a regulation of the conduct of servants in general.[16]

Before the end of the seventeenth century corporal punishment was extended to cover offenses against the dignity and status of the master.

> In 1673 the General Court ordered that a servant "for scandalous false and abusive language against his master have thirty-nine lashes publicly and well laid on in James City and that he appear at Middlesex County Court next and there openly upon his knees in the said court ask forgiveness which being done is to take any further punishment allotted him."[17]

Control by the master over the servant's person and liberty of action was given after a plot of servants against their masters in 1663 was discovered. The great alarm led to the strict regulation of such liberties as leaving the master's plantation and assembling together. After 1726 further powers of control over white servants by masters ceased and the institution steadily declined in the face of rising Negro slavery.[18]

Negro Slavery and Its Control

The transition from trade to agriculture evidently brought a new type of settler to Virginia. In 1619 John Pory, Cambridge graduate and gentleman, lamented that "in these five months of my continuance here there have come at one time or another eleven sails of ships into this river; but freighted more with ignorance, than with any other merchandize." Ten years earlier the ships had brought well-born adventurers; now they brought those who intended to grow tobacco.

> Henceforth the future of the colony was with those who could clear the forest, establish plantations, and withstand the agues of the mosquito-infested lowlands. The leaves of fate for Virginia were not to be thumbed in a book. They stood broad and strong over the rich bottom-lands, where the summer sun seemed to the onlooker to deck their oily surfaces with a coat of silver. In the days of the gentlemen adventurers nine men wrote about the history of the colony; in the days of the tobacco growers a century could not show as many.[19]

The first generation born in Virginia grew up in a rough world whose virtues were not those of niceness but of resourcefulness and enterprise. They were uncouth and iconoclastic. The existence of a higher and lower class was recognized, but class lines were not drawn too hard and fast; the requirements of common work together took care of that. But most of the settlers were drawn from the same English national culture, with the same religion and the same language. In 1619, to the extent that equality of competition prevailed among the new class of tobacco growers on the Virginia frontier, social relations between them must have been fairly democratic. All might hope to become "gentlemen" because all might hope to possess land, and in England the possession of land had carried that coveted status.

This was the situation when "about the last of August (1619) came in a Dutch man of Warre that sold us twenty negars." With this sentence in his journal John Rolfe notes the entrance of a new group different in skin color, language, and religion. The Negro stranger must have been regarded as a queer curiosity who, because of his skin color and cultural heritage, could not conform to the English traditions if he would. As long as the planters were able to maintain their Old World traditions and not revert too far toward Indian methods of utilizing the land where both Indians and Negroes might meet them on terms of competitive equality, the Negro necessarily remained to some degree an unassimilated and foreign element. So long as this Old World culture could be maintained the Negro was naturally placed at a competitive disadvantage greater than that of the white indentured servant who shared the cultural traditions of his master.

Nevertheless, in 1619 there was little in the *mores* of the English planters which countenanced actual slavery, and the twenty Negroes took their place in the colony as indentured servants at that time. There was initially some difference in the treatment of white and Negro servants, but, on the whole, there was a tendency to regard all servants, white and Negro, as members of a lower class with similar relations toward members of the higher class, the planters. Indeed, there seems to have been a good deal of sympathy and intimacy between all servants, out of which came sexual unions, and joint revolts against the planters. However, the competitive opportunities of the Negro were restricted with greater ease and acceleration than in the case of the white servants. His relations with the white master to whom he was indentured and who tended to regard him with a certain sense of responsibility, or even potentially competitive[ly]. It was a feudal relationship; how legal processes operated to inject a more objective property attitude into this relation will be noticed later. To the extent that both white planter and Negro servant acquiesced in this relationship—his need for labor made it easy for the planter to do so—there was finally built up a conception by each group of the other as members of a separate and distinct "race," a different species, each biologically equipped with immutable instincts and dispositions constituting the one into a "higher" and the other into a "lower" race. By this time the white servant was practically eliminated from the plantation, which then became an organization of races rather than one of classes.

Economically these were relations whereby the planter got laborers to grow his tobacco and the Negro got someone who was responsible for his sustenance as a minimum consideration. It was the sort of labor necessary and desirable for tobacco-growing under frontier conditions. It assured the planter a more or less long-time stable supply [of labor]. Indentured servitude provided labor for a period of years only, but slavery made it perpetual from generation to generation. One aspect of the process of enslavement was, therefore, the lengthening of the period of servitude. In addition to being cheap and stable in time, labor for tobacco production on the frontier had necessarily to be mobile. In the absence of cheap fertilizer tobacco planters found it necessary after a few years of cultivation of particular tracts of land to move to virgin fertile soil. The plantation in the South was an efficient device for skimming the fertility of the land and the possibility of unrestricted expansion was an essential requirement.[20] As opposed to serfdom the incidents through which indentured servitude and slavery developed were those designed primarily to secure labor mobility. These forms of forced labor reproduced, or tended to reproduce, the personal rights and obligations of feudalism with the laborer's right of tenure left out. It was the personal attachment to a lord, wherever he might migrate, rather than to the lord plus the land, that characterized slavery.[21]

On the social and psychological side, the relations which eventuated in slavery grew out of a community of experience in the production of a crop on an isolated plantation. The colony as a whole, regarding the Negro as an abstraction, as the South still does, might be fearful of his presence, especially as he increased in numbers, but on the plantation frontier where master and man were brought into daily relations and confidence established, the relationship in all probability became a comfortable one for both persons.[22] It was a family relation; the master felt secure and had a sense of added dignity because he had the permanent assistance of his man. Common law marriages, under conditions where there were few to oppose them, were not infrequent between masters and white or Negro women servants.[23]

The servant, white or Negro, was usually a familyless individual. He was an adult when he came to Virginia and began to serve out an indenture for perhaps five years. Five years of servitude in the isolation of the rural plantation would naturally build up habits of work and attitudes of subordination which would be pretty well ingrained when freedom day came. If, for some delinquency, the servant's period of servitude was extended for a few years longer, he would find it more difficult to break off old relations and go out "on his own." Perhaps he was no longer a young man, and ageing years rendered him even more dependent upon the master and his family. We are familiar with the familyless man or woman in many present-day communities who attach themselves to a family from whom they receive support and affection.[24] The relationship in indentured servitude lent itself to the building up of "moral insolvency" among the servants who might easily be led to depend upon the maintenance and protection of a master and to feel less responsibility for themselves. This is a human attitude which is easily understood, not unlike the religious attitude of surrender and the willingness to do the bidding of the Lord. In such a situation what is demanded of a servant, white or Negro, is not merely service but loyalty.[25] Thus the master and servant relationship, imbedded as it was in attitude and custom, and reinforced by the economic need for labor, tended to lengthen itself into additional years without necessarily calling in the power of the law to achieve that end. In the case of Negro servants, where social dependence might be expected to be stronger, the lengthening of the relationship would follow with less difficulty. Such attitudes would also aid in the settlement of new plantations on the frontier in which case the situation would be analogous to that of any family settling in the forest, making a clearing, and planting a crop. The forest might seem to invite escape but no one thinks of escaping, for no one broods over his fate, and the fact of bondage is not reflected upon. For this reason the stratified slave plantation might move westward and settle the land along the frontier as a community as well as an industrial agricultural establishment.

The original cleavage between the Negro and white settlers of Virginia, so far as there was any, seems to have been religious rather than racial.[26] Early church and court records in Virginia reported them as "infidels" or "heathen" whether they were baptised or not.[27] Until about 1682 religious differences rationalized all discriminations made against Negroes, but it was a half century before those discriminations reached the stage of legal slavery.

In his *A History of Slavery in Virginia,* Ballagh was first in pointing out the error in the assumption that Negro slavery was introduced full-grown into Virginia. The twenty Negroes landed at Jamestown in 1619 were obtained by exchanging public provisions, "and they were put to work upon public lands to support the governor and other officers of the government; or, as were several in Virginia, they were put into the hands of representative planters closely connected with the government in order to separate them from one another." In 1623 they were living in seven different settlements.[28] The masters of these Negroes were possessed of the right to their services but not to their persons. The twenty-three Negroes, living in Virginia in 1623, were all listed as "servants." "In the records of the county courts dating from 1632 to 1661," says Russell, "negroes are designated as 'servants,' 'negro servants,' or simply as 'negroes,' but never in the records which we have examined were they termed 'slaves.'"[29]

A Negro named Anthony Johnson was probably a member of the first group introduced into the colony. In 1652 the county court of Northampton County heard and granted his petition asking to be excused from the payment of taxes for public use because of loss of property by fire. The year before Johnson had been assigned two hundred and fifty acres of land in fee simple. It is probable that he had long been a free man established in the midst of his own acres.

Johnson was not only a free man but he himself became the master of servants. The year after he had been freed from the payment of taxes he again had occasion to come into court. John Casor, a Negro who came to Virginia about 1640, complained that although he had been indentured for seven or eight years he was still kept in servitude by his master "for seaven years longer than hee should or ought." Now John Casor's master was none other than Anthony Johnson. Both were Negroes. An act of 1670 forbade free Negroes to own Christian servants but recognized their right to own servants of their own race. The court's decision in this case is interesting and significant.

Casor appealed to Captain Samuel Goldsmith to see that he was accorded his rights. Goldsmith demanded of Johnson the servant negro's indenture and was told by Johnson that the latter had never seen any indenture, and "yt hee had ye Negro for his life." Casor stood firmly by his assertion that when he came in he had an indenture, and Messrs. Robert and George Parker confirmed his declaration, saying that "they knew that ye sd Negro had an Indenture in one Mr. (Sandys) hand, on ye other side of ye Baye & . . . If the sd Anth. Johnson did not let ye negro go free the said negro Jno. Casor would recover most or his Cows from him ye sd Johnson" in compensation for service rendered which was not due. Whereupon Anthony Johnson "was in a great feare," and his "sonne in Law, his wife, & his own two sonnes persuaded the old negro Anth. Johnson to set the sd Jno. Casor free."

The case would be interesting enough and very instructive if it had ended here, but the sequel is more interesting still. Upon more mature deliberation Anthony Johnson determined to make complaint in court "against Mr. Robert Parker that hee detayneth one Jno. Casor a negro the plaintiff's Serv(an)t under pretense yt the sd Jno. Casor is a freeman." His complaint was received, and the court, "seriously considering & weighing ye premises," rendered the following verdict, than which there are none stranger on record: "The court . . . doe fynd that ye sd Mr. Robert Parker most uprightly keepeth ye sd Negro John Casor from his r(igh)t Mayster Anthony Johnson & . . . Be it therefore ye Judgment of ye court & ordered that [t]he sd Jno. Casor negro shall forthwith return into ye service of his sd Mayster Anthony Johnson and the sd. Mr. Robert Parker make payment of all charges in the suite and execution."[30]

There is no better evidence than is afforded in this case for the statement that slavery was an undesigned, unpremeditated response to the needs of agricultural industry on a thinly settled frontier where it was easily possible for men to live without offering to work for wages. This case, one of the earliest showing the establishment of lifetime servitude, is all the more interesting because the two principals were themselves Negroes.[31] "Between 1640 and 1660," according to Russell, "slavery was fast becoming an established fact. In this twenty years the colored population was divided, part being servants and part being slaves, and some who were servants defended themselves with increasing difficulty from the encroachments of slavery."[32] The tendency toward dual standards of punishment for offenses committed by white and Negro servants is evident from a case in 1640 in which two white servants and a Negro servant ran away from their master, were captured and brought back for trial. All three were sentenced to a whipping and to have thirty stripes each. In addition the two white servants were ordered to serve their master an additional year and the colony for three years after their regular terms of service had expired. But "the third, being a negro . . . shall serve his said master or assigns for the time of his natural life."[33]

According to Russell, our best authority for these facts, few Negroes brought into the colony after 1640 became free, and for fifteen years before the passage of the first act of the Virginia Slave Code in 1662 they were being sold and recorded as lifetime servants. However, nothing was done to reduce the recognized free Negro to slavery; on the contrary the laws which made explicit reference to slavery after 1662 guaranteed his status and recognized his right to hold servants of his own race. But lifetime servitude for Negroes had been established as a fact. "They are call'd Slaves," said Beverly, "in respect of the time of their Servitude, because it is for life."[34] They were servants against whom the usual punishments by addition of time could not hold, for they were "incapable of making satisfaction by addition of time," according to an act of 1651.[35]

Servitude for life, however, was not yet legal slavery. The first Virginia act to recognize what had already become a fact in custom was an act of 1662 to settle questions concerning the status of the children of servants for life, probably all of whom were Negroes. "It was evident," says Ballagh, "that parents under an obligation of life service could make no valid provision for the support of their offspring, and that a just title to the service of the child might rest on the master's maintenance, a principle which was later commonly applied to cases of bastardy in servitude."[36] The act determined that the child should follow the condition of the mother. It was this "principle of heredity which differentiated the condition of slavery from the condition of servitude. But even the act of 1662, Ballagh maintains, was not intended to create a race of slaves but to prevent sexual irregularities and race mixture. Hereditary slavery grew directly out of the problem of the mulatto:

> Notwithstanding its effect it is clear that the purpose of the act of 1662 was primarily punitory. It was designed to prevent race mixture rather than to create slaves. The "spurious issue," as it was termed of whites and blacks was at all times abhorred. In the earliest instances of fornication with negroes, in 1630 and 1640, the severest penalties were inflicted. Whipping and public confession, were exacted of both the offenders in 1640. An additional penalty was imposed upon the female in 1662 of the bondage of her issue, which it was hoped would effectually check the evil. Probably little trouble from the growth of mulattoes was actually experienced until the second half of the century, when both negro and Indian population had greatly increased. The name "mollatoes," of Spanish-American origin, first occurs in an act of 1682, applying only to a class of imported cross-breeds, but by 1691 its extension to a native element seems to have been established. At this time negro and Indian bastards were increasing, and the offense of race mingling had extended even to white women. Thus arose a new difficulty in the clear probability of a class of free mulattoes, but the manner in which the question was disposed of shows conclusively that prevention of an "abominable mixture" and not enslavement was the end in view.[37]

The problem of interracial sexual contacts and illegitimacy was one of the most stubborn to confront the colony. Although the white planter class was not free from such contacts, the greatest amount of irregularity seems to have characterized the relations between white and Negro servants. These probably began shortly after the introduction of the first Negros, for in 1630 the court ordered a white man punished for "abusing himself to the dishonor of God and the shame of Christians by defiling his body in lying with a negro."[38] From this time on the General Assembly, the courts, and the church wrestled with the problem, but in spite of everything the country, according to Peter Fountaine, swarmed with mulatto bastards, who, "if but three generations removed from the black father or mother, may, by the indulgence of the laws of the country, intermarry with white people, and actually do every day so marry."[39] Toward such relations between Negro and white servants the planter class expressed such judgments as "disgrace of the nation," "shameless matches," "defiling the body," "unnatural unions," etc. They point to a situation where whites and Negroes worked side by side on the plantations for a common master.[40] Together on the plantation engaged in the same sort of manual labor there was little antipathy between the white servant and the negro servant or slave. But events were in process which were to separate and institute competitive relations between them resulting in the well-known hostility between "niggers" and "poor white trash." These events were such as greatly to alter the numerical proportions of the races.

After 1619 the Negro population grew slowly; five years after their introduction there were only twenty-one. Even in 1670 they constituted only five per cent of the total population of the colony. The tobacco trade, which had suffered depression for a quarter century before 1682, now suddenly expanded very rapidly. But unlike earlier periods of prosperity the profits of the expanding trade did not greatly stimulate the growth of the number of small planters and farmers.[41] This is explained by the fact that the new prosperity came not from a rise in price but from a lowered cost of production. Small planters and farmers, or servants looking forward to the expiration of their period of servitude, could not command the capital necessary to introduce the new economics and profit at a price for their tobacco which otherwise proved unremunerative. Only the large planter with capital was benefited by the expanding market, for he was able to stock his plantation with a new source of human energy obtained from African slave traders and which was cheaper than the old source provided by indentured servitude. The planter could now purchase lifetime slaves directly from the slave trader, for the necessary customs and laws had now evolved in Virginia to support slavery. From 1680 on, therefore, Negro slaves were imported in rapidly increasing numbers and purchased by the holders of large estates. In 1730, out of a total population of 114,000 in the colony, 30,000, or 26 per cent, were Negroes. White servants were gradually crowded off the plantation by Negro slaves until only the white overseer was left. They drifted into the infertile spaces between the plantations or squatted along the further Virginia frontier before they finally spilled over into Kentucky or formed the individualistic Southern element without capital which migrated into the Ohio Valley or into the deep South.

The racial basis of slavery was brought directly to the fore as the plantation became an organization of Negro laborers and white supervisors. When it was found necessary to justify slavery, however, resort was had to the prevailing conception of Europeans that it was right for members of a Christian nation to hold inhabitants of a heathen nation in bondage since slavery was a means of bringing men to Christ. Negroes might be enslaved not because they were Negroes but because they were not Christians. This vew was accepted by Virginia planters in the seventeenth century, but it brought the unexpected and troublesome problem of the converted slave. A rationalization justifying initial enslavement and the slave trade failed to justify holding a slave after he became a Christian. To avoid loss of their property planters strongly opposed the efforts of individuals and societies seeking the Christianization of negro slaves. It became necessary for the Virginia Assembly to declare that conversion did not entail freedom, but this was clearly inconsistent with the prevailing ideology upon which slavery was based. A new justification was necessary.

It was found in the idea of race. The conception that a certain group of men coming from the same general territory and possessing similar physical marks were innately and immutably different and constituted a separate species powerfully reinforced the position of slavery in Virginia and was formally incorporated in the law of slavery by the Assembly. Such a conception permitted the Christianization of slaves but did not interfere with their bondage. It regulated competition in the interest of the planter aristocracy and permitted men to live and work together on the edge of the wilderness within the structure of a single institution instead of scattering over a larger territory to settle as squatters. The idea of race, and of slavery based upon race, was thoroughly accepted by both the master and the slave classes as the norm of relations. If the legal sanction of slavery has been destroyed the idea of permanent racial differences continues as a powerful form of control within the plantation institution, powerful because it is taken for granted by both groups. The plantation henceforth was an organization of race relations.

The Evolution of the Planter

Virginia's plantation aristocracy, the planter class, evolved simultaneously with the changes that brought indentured servitude and slavery. When, in 1619, the Virginia Company made a division of lands among their "adventurers of the person," or planters, those who had arrived in the colony before 1616, and who therefore had endured the greatest hardships, were each given 100 acres of land per share of stock. These were called "ancient planters." Those who came after 1616 were given only fifty acres per share of stock.

Here we have a hint as to the probable reason the term "planter," which originally meant simply one planted or settled, came to gain prestige value as the title of the plantation aristocracy. There seems to be in human nature a general disposition to accord distinction to first things. Every group tends to remember its pioneers. In his essay on Provincialism, Josiah Royce says that out of the extreme democracy of California frontier days there emerged a semblance of an aristocracy whose prestige rested on the fact that its members were the first to arrive.

It must have been similar in early Virginia. Those who pioneered at Jamestown gained experiences and rights which entitled them to respect and made them "ancient" planters in the eyes of those who came later. For, of course, those who came later, especially the indentured servants, crossed the ocean under very different circumstances. These later arrivals were immigrants, not planters, in the original sense of that term. When tobacco cultivation began, the connection of "planter" with agriculture was easily made. When it became necessary to distinguish between the holder of indentured servants and slaves, and those free white colonists who did not hold servants or slaves, there was a place in the language for both planter and farmer to fill the needs of ordinary conversation. In short, planter was used not only to designate a vocation but it became also a status to be achieved. When slavery took the place of indentured servitude, the status of the planter was elevated and lengthened into a permanent relation.[42]

The strength of the planters consisted largely in the fact that they either possessed capital of their own or were able to command it. After 1610 few of the planter-adventurers of John Smith's day were left in the colony.[43] When the profits from tobacco were discovered a new class of settlers began to arrive. Most of these were indentured servants but many were freemen with capital; a good many, if not most, of the latter were merchants or the sons of merchants driven from England by the civil wars. Wertenbaker says:

> Beyond doubt the most numerous section of the Virginia aristocracy was derived from the English merchant class. It was the opportunity of amassing wealth by the cultivation of tobacco that caused great numbers of these men to settle in the Old Dominion. Many had been dealers in the plant in England, receiving it in their warehouses and disposing of it to retailers. . . .
>
> The life of the Virginia planters offered an inviting spectacle to the English merchant. He could but look with envious eyes upon the large profits which for so many years the cultivation of tobacco afforded. He held, in common with all Englishmen, the passion for land, and in Virginia land could be had almost for the asking. He understood fully that could he resolve to leave his native country a position of political power and social supremacy awaited him in the colony.[44]

These merchant-planters possessed the capital necessary for the initial expense which made the plantation capitalistic in character. If the plantation was a large one the initial capital outlay might easily amount to a considerable sum. In 1732, Robert Beverley spent £11,234 17. 8½ in taking up a tract of 24,000 acres and about a third of this amount went for clearing a portion of the estate.[45] In 1690, William Fitzhugh advised anyone planning to raise tobacco in Virginia to deposit 150 or 200 in the hands of a London merchant with which to buy lands in the colony, and about the same amount with some member of the Royal African Company for slaves.[46] Of those who were able to raise such sums and had the courage and ability to supervise their investments in Virginia, some were men of social rank in England but probably most were not. The essential thing in early Virginia was not social rank but capital, and when this was acquired an Englishmen of whatever social position had no difficulty in acquiring the estate of a planter. Thus:

> Samuel Mathews, a man of plain extraction, although well connected by marriage, was a leader of the colony. In political affairs his influence was second to none, and in the Commonwealth period he became governor. He is described as "an old planter of above 30 years standing. . . . He hath a fine house, and all things answerable to it; he sows yearly store of hemp and flax and causes it to be spun; he keeps weavers and hath a tan house . . . hath 40 negro servants, brings them up to trade, in his house; he yearly sows abundance of wheat, barley, etc." . . . Adam Thoroughgood, although he came to Virginia as a servant or apprentice, became wealthy and powerful. His estates were of great extent. . . . Captain Ralph Hamor, a leading planter in the days of the Company, was the son of a merchant tailor. Thomas Burbage, was another merchant that acquired large property in Virginia and became recognized as a man of influence. Ralph Warnet, who is described as a "merchant," died in 1630, leaving a large fortune.[47]

Virginia's merchant-planters of the seventeenth century were apparently men of very different characteristics from the planters of the eighteenth century. They had not had time to acquire the habits and the traditions of the new order or to break fully away from their mercantile instincts. The Dutch trader, De Vries, said of them, "You must look out when you trade with them, for if they can deceive any one they account it a Roman action."[48] Wertenbaker writes in some detail of their "extraordinary pride and ambition," a phrase used by Governor Nicholson to describe Robert Carter, whose arrogance earned him the title "King" Carter.

> But as time went on a decided change took place in the nature of the Virginian's pride. During the 18th century he gradually lost that arrogance that had been so characteristic of him in the age of Nicholson and Spotswood. At the time of the Revolution are found no longer men that do not hesitate to trample under foot the rights of others as Curtis, Byrd, and Carter had done. Nothing could be more foreign to the nature of Washington or Jefferson than the haughtiness of the typical Virginia planter of an earlier period. But it was arrogance only that had been lost, not self-respect or dignity.[49]

The change in the personality of the planter accompanied the complete establishment of his division of labor and the fixation of the habits that went with that division of labor. Isolated upon his large plantation and occupied with the cultivation of tobacco and the supervision of servants and slaves, it was inevitable that he should develop the attitudes and bearing of the aristocrat. His was a division of labor which forced him to keep in touch with world affairs. Naturally he was interested in the market for his tobacco, and that market was in England. Hence his close interest in European affairs, his correspondence with English factors, the education of his children in England, and his acquaintance with English newspapers and books. The close interest of the Southern planter in the English market and in English culture was to a large extent a fact until the Civil War. In addition, the planter was the most mobile member of the plantation and his was a mobile division of labor. The contacts thus established and maintained by the planter with other planters contributed to common understandings and a knowledge of their common interests which helped to organize them in their control of the state.

If the Virginia planter could not be an absentee in England like the wealthy West Indies planter, he might, if he was successful, import into Virginia the best that England had to offer. The furnishings for their homes, the latest fashion creations, fine carriages, carpets, and many other luxuries were purchased through English factors. Thomas Mann imported both materials and workmen to build his mansion, which was five years under construction.[50]

Thus by the time of the Revolution Virginia had developed from within a well-entrenched landed aristocracy. The large plantation systems, the wealth of the planters, the management of indentured or slave labor, leisure time and direct connections with the English market all resulted in establishing that type of eighteenth-century Virginia planter, represented by men such as Washington, Jefferson, Madison, and Monroe, which is thoroughly familiar in American tradition. However, the reaction against the exploitive and wasteful methods of cultivation identified with the planter caused some estate holders, including Washington, to prefer the title "farmer" since they counted themselves followers of the most progressive cultivators of England and Europe.[51] With the decline of the Virginia plantation and the trend toward the small farm, planters became somewhat fewer and less important. But so long as there were plantations there were planters. When J. D. B. DeBow, of Louisiana, became superintendent of the United States census in 1850, he undertook to enumerate separately the planters and farmers of the Southern states. The definition of a planter used by the census of that year is not given but, whatever it was, Virginia was found to have 1,374 planters and 106,807 farmers. The census of 1860, however, found only 80 planters in the state as against 108,958 farmers.

The Humanization of the Plantation

After "plantation" and "colony" had become differentiated aspects of the Virginia community,[52] the form of government designed for the plantation was conceived in terms of the forms and practices of the English manor. To Captain Martin, for example, was granted a particular plantation to possess "in as ample manner as any lord of any manor in England." Martin seems to have exercised his powers by refusing to submit to the laws of the general colonial government. The Colonial Assembly in turn refused to seat his delegates until he had surrendered his patent and accepted a new one with restricted powers. Thus plantation and colony did not become differentiated without some conflict of interests. This case is important because it shows that in the absence of a strong centralized colonial government, planters on isolated plantations tended to assume considerable powers which custom and finally the General Assembly itself recognized.[53]

Old moulds of government may be instituted in a new settlement, but government itself is not so instituted. It arises out of the specific circumstances under which men meet and deal with each other. Such circumstances were different in manorial England and in Virginia. The plantation in Virginia sought not rent from tenants but profits from tobacco, and indentured servants and then Negroes, who were strangers to English traditions, were imported to grow it. Difficulties of transportation and communication effected an extreme localization of life; where these are restricted we have no reason to expect anything other than highly decentralized social, economic, and political units. Thus the individual plantation was left very much to impose its own control in terms of its own economic purposes. Slavery, corporal punishment, and even the power of life and death were aspects of this control.

We have seen how the Virginia Assembly in 1662 recognized the right of the individual master to administer corporal punishment to indentured servants, a practice which probably already had long been exercised by masters.

> Until 1723, if a slave chanced to die under, or in consequence of lawful correction it was viewed as merely a lamentable and "accidental homicide." An act of that year declared such killing of a slave to be manslaughter only, and not liable to prosecution or punishment. But if a single credible witness affirmed before the county court that the slave was killed "willfully, maliciously, or designedly," the perpetrator might be indicted, and, if convicted, punished as a murderer.[54]

At least before 1669 the master's power of life and death over slaves was practically absolute as far as the law of the colony was concerned.

The type of organization in which authority resides depends upon the character and extent of the interaction of individuals upon each other, upon the solidarity of interests—ultimately upon the means and extent of communication. In early Rome, the house-father had legal authority to sell his children and to inflict capital punishment on them. His wife and even the grownup son and the latter's children were under his power.[55] The status of the family was very different then from what it now is, and powers which are now delegated to the larger community were then possessed by the head of the household. The *familia* was itself a political institution and as such held the power of life and death; the state, which assumed these powers as it became more integrated, still holds them.[56]

We may suppose that the planter's power of life and death developed in an isolated situation in very much the same way and for very much the same reasons. If among his slaves were some who were his own children, a parallel with the *potestas* of the Roman house-father would be more closely approximated. Until the colonial or state government became sufficiently integrated and centralized to permit it to assume this power unto itself alone, neighbors might condemn the taking of the life of a slave, but as a rule they would feel it more expedient not to interfere, just as parents pursue that policy with reference to other parents' children. But a strong public opinion could and did arise to limit the master's power of life and death.

Slavery, and the forms of discipline that went with it, grew up in Virginia as a set of customary relations to meet the needs and exigencies of agricultural production. But it grew up within the legal system of another social order, England. Contact with this legal system effected a radical transformation in the very nature of slavery. Apprenticeship and indentured servitude were legally sanctioned and enforced means of regulating labor. One of the earliest legal questions in connection with indentured servitude in Virginia had to do with the right of the master to assign his servant's contract whether the servant gave his consent or not. The courts recognized this as a right of the master, and his control came to be defined as a property right transmissible by sale and inheritance. The individual who bought or inherited the slave as property would, of course, rarely maintain the personal relations in which the slavery developed and was embedded. It was in the slave trade that the relation tended most of all to become an abstraction divested of all its human associations, restraints and inhibitions. In this form it was recognized in the law and maintained by public authority. Hence it was in the slave trade that the most ruthless aspects of the system came to the fore, where, indeed, "ruthlessness was the law." Even the intimate relations of the slave family could not withstand the separating effects of the slave trade.[57]

In an economic order such as that of antebellum Virginia where the demand for cheap labor to produce tobacco was greater than the supply, the slave trade provided a means of quickly distributing illiterate workers where they were most needed. It offered the planter a way of obtaining labor for routine work in the same manner that he obtained his livestock.

Throughout its annual cycle of cultivation tobacco was a crop which lent itself admirably to routine labor. First sown like lettuce in beds the plants were transplanted early in May in hills about four feet apart. Constant attention was required. Fields must be kept clear of weeds, and plants must not be permitted to have too many leaves; after they had been set out about a month slaves were put to work "topping" them. Once topped, the plants could grow no larger but could still put out "suckers" between the leaves. These had to be removed once a week until the leaves ripened in July or August. Between toppings the weeds had to be watched, the ground kept in proper condition by hoeing, and the tobacco worms taken off the plants. When in August the leaves began to thicken and spot and turn from green to brown, a new kind of work began. As fast as the plants reached the cropping stage they were cut down, heaped on the ground, and allowed to sweat over night. Later the slaves carried the plants to the tobacco houses on their shoulders or dragged them in on the branches of trees. After sweating for awhile in the tobacco houses the leaves were "stemmed" and graded for the market. Finally came the task of packing them in hogsheads which were rolled to the wharves or warehouses.[58]

Each of these steps in the cultivation of tobacco demanded manual labor, but not very skilled labor. Prior to 1860 American agriculture was carried on largely by hand, and Virginia agriculture was no exception to the general rule. Machine appliances were relatively unimportant. Wooden plough tips and shares did not last long in newly cleared lands and iron ones were both expensive and hard to keep in repair in a country in which there were not many craftsmen. Consequently ploughs were primitive and clumsy, sometimes only grubbing hoes tied to a plow beam. Moreover they were usually of English manufacture and ill-adapted to the nature of the land.

Both machinery and industrial slavery are power systems—systems for utilizing energy—and the economic unit tends to get adjusted to the type of power available, whether machine power or human power. The farm tractor, first thought of as a machine which would take the place of the horse, proved to be no mere substitute for animal power but served to pitch agriculture in the areas where it is used upon a new level and to reorganize the scale of that agriculture in new dimensions. In the absence of machine power on the Virginia frontier, large-scale power could be generated only by multiplying the number of human hands. This was impossible on the tobacco frontier except by some form of forced labor, such as slavery, supplied by the slave trader. Capital invested in manpower gave the plantation exactly the same advantage over the small farm as present-day "power farming" with capital invested in machines gives over the smaller unit. The Virginia plantation represented "power farming" under circumstances where more men rather than machines furnished the added power.

If plantation slavery as an industrial system tilted the scales of human relations and led to inhuman judgments and cruel actions, there were counteracting forces which, in certain ways, tilted them the other way. The background of plantation relations had much in common with the foreground of our contemporary experiences, for human nature changes little even while it is adapting itself to a wide variety of conditions. The economic and political systems in which men live are of great consequence to their actions but these systems do not determine relations absolutely. Love and hatred, confidence and jealousy, understanding and distrust are common human qualities which crop up in convents and penitentiaries, missions and universities, congregations and pirate-crews. Murderers and slave dealers often show extraordinary generosity and, according to an old story, Satan once raised an insurrection in heaven.

Opposed to the disposition to regard African slaves as so much livestock to be domesticated for plantation labor, a very human, but possibly economically unwise, process was going on. It was a change whereby slave livestock, purchased or inherited, was in process of becoming persons, and of gaining the rights and privileges of persons. Men who must cultivate a garden together, even when some are slaves and others masters, have a way of getting under the skins and making demands upon the sympathies of each other. Even conflicts establish the inter-human ties more closely. Regardless of his views concerning that abstract category "the Negro," the planter became closely attached to individual Negroes, not because they were better than ordinary, but simply because he knew them. As against "the Negro" whom he feared, he knew and trusted "Uncle Bob" and "Aunt Emma." The truth is that when planters began to look upon their slaves as human they themselves were no longer free; they could not disregard human claims and attachments.

It was for this reason that, along with the process establishing slavery in Virginia, went another which led to freedom. The sale or inheritance of Negroes without the personal relations in which their slavery was embedded resulted in that ruthlessly abstract conception of the law which we have noted. To avoid the harsh consequences which might fall to their slaves if they came into the possession of owners not recognizing such personal ties, many planters manumitted them. A planter who had no scruples against owning slaves himself, because he knew his own sentiments toward them, might regard their ownership by someone else with much aversion. When he came to make his will he would have occasion to reflect seriously upon his relations with life-long companions and to be moved to give them their freedom with property. In documents disposing of property and designating heirs may be read, between the formal legal lines, the sentiments of men whose hard matter-of-fact exteriors in daily life might seem to mark them as individuals without sentiment. Thus Thomas Whitehead of York County, Virginia, sometime between 1657 and 1662, by will emancipated his slave, John, and bequeathed to him a great variety of clothing, and also two cows, ordering that he should be allowed the use of as much ground as he could cultivate, and the possession of a house. So great was his confidence in the discretion and integrity of this negro, that he appointed him the guardian of Mary Rogers, a ward of Whitehead's, and overseer or her property, offices which the court refused to suffer him to fill.[59]

The following is an extract from the will of John Warwick of Amherst County, probated in 1848:

> I, John Warwick, of the County of Amherst, . . . do make, publish and declare this my last true will and testament. . . .
>
> First: The future condition of my slaves has long been a subject of anxious concern with me, and it is my deliberate intention, wish, and desire that the whole of them be manumitted and set free as soon after my demise as the growing crops shall be safe and the annual hires terminated, not later than the end of the year of my death, to be removed, or so many of them as I do not manumit and send to a free state during my life, with the exception hereinafter named, and settled in one or more of the free states of this Union under the care and direction of my executors, hereinafter appointed. Indiana is my choice.
>
> Second: To carry out the above bequest . . . next to the payment of any debts I may owe, my funeral expenses, and the charges of administration of my estate, I hereby declare that it is my wish and intention that my slaves shall on being emancipated have the whole of my estate now in being, or hereafter to be acquired, . . . for the purpose of creating a suitable fund in the hands of my executors for their comfortable clothing, outfit, travelling expenses and settlement in their new homes.[60]

Such actions were by no means exceptions. How numerous manumissions must have been may be judged from the fact that in 1860 Virginia had 55,269 free Negroes which, however, included the children of free Negroes who never knew actual slavery. This represented slightly over nine per cent of Virginia's entire Negro population. When it is compared with Arkansas' 144 free Negroes out of a total Negro population of 111,259 it will be realized how far the frontier had passed away in Virginia as compared with plantation communities in the Southwest.[61]

The presence of such a large number of free Negroes in a social order based upon Negro slavery was an indication of the disintegration of the Virginia plantation system from within. They constituted a population element with no

88

defined place in that system and were regarded by white public opinion as a constant threat to it. Until 1860 the legislature and courts of Virginia wrestled with the problem of the free Negro without success. A few were colonized in Liberia and more were sent to the free states of the North and Middle West. Legislative actions directed toward regulating both slaves and free Negroes were approved by public opinion but nullified by human relations with particular slaves and free Negroes. As a writer quoted by Russell expressed it:

> As Legislators, impressed with the jeopardy that threatens the public safety, men readily give their assent to any measure that seems calculated to protect it, but when they return to the bosom of their families and are surrounded by those among whom they were born and nursed and from whose labor they obtain the means of comfort and independence the sentiments of the legislator are frequently lost in the feelings of humanity and affection in the private man.[62]

For the same reason it was impossible to enforce laws which required the removal of free Negroes from Virginia. Not only did human ties and sentiments prevent the operation of such laws but neighbors were unwilling to part with the services of the free Negro who occupied a division of labor or had acquired skills useful to the community. And so Virginia, in spite of her fears, continued to multiply her free Negroes and to maintain them within her borders.

They were symptomatic of the profound changes which the plantation was undergoing. Human relations were in process of transforming the estate; it was becoming merely a large family. When Thomas Massie, spendthrift planter of Virginia, became heavily involved in financial difficulties, he proposed in a letter to his father to diminish his obligations by selling his highly improved homestead in order to retain his slaves. He wrote:

> To know that my little family, white and black, [is] to be fixed permanently together, would be as near that thing happiness as I never expect to get. . . . I had rather leave this fine place and go to a much humbler one than to part with my negroes. Elizabeth [his wife] has raised and taught most of them, and, having no children, like every other woman under like circumstances, has tender feelings toward them.

The reply of his more successful father, William Massie, illustrates the other side of the plantation with which these familial sentiments were in conflict. Like a good efficiency engineer, the elder Massie advised his son to sell his slaves instead.

> Indeed, my son, it seems a blind infatuation that you should be so devoted to that kind of property which must have been greatly instrumental in all

your embarrassments. . . . Negroes won't do unless they are made profitable, which yours have not been or you would have made money instead of being ruined. Negroes should be made to live comfortably, not luxuriously and tenderly, be furnished with the necessaries not superfluities of life, made to work on some crop that will pay and not on buildings and projects; and then they are most generally profitable if the product of their labor is not wasted—but not always.[63]

"—but not always." Changing economic factors as well as humanizing influences were breaking down the plantation and paving the way for the succession of a new economic order.

5

The Plantation and the Frontier

Economic Changes and the Small Farm in Virginia

In the days when the tobacco fever was at its height, about 1720, Beverly, the historian of Virginia, called attention to the almost absolute dominance of the staple and the consequent lack of adequate provisions for food.

> What advantages do they see the neighboring Plantations make of their Grain and Provisions, while they, who can produce them infinitely better, not only neglect the making a trade thereof, but even a necessary Provision against an accidental scarcity, contenting themselves with a Supply of Food from hand to Mouth, so that if it should please God to send them an unseasonable Year, there would not be found in the Country Provision sufficient to support the People for the Three Months extraordinary.[1]

In 1860, according to Craven, the two states of Virginia and Maryland "had come largely to the small farm and the small farmer," and "extensive crops and methods had given way to intensive cultivation or diversified production."[2] Virginia agriculture and the Virginia plantation had undergone great changes since Beverly's day.

We have seen how the price of tobacco fell from three shillings a pound in 1619 to an average of about three pence a pound in the decade before the Revolution, while production was steadily increasing. The devastation of plantations and the loss of markets during the Revolution resulted in the almost complete disappearance of tobacco, but after about 1783 production rapidly increased and Virginia was soon exporting more than before the Revolution. Nevertheless, changes were in process which were to relegate tobacco to a secondary position in Virginia's agriculture. From [a high of] 118,460 hogsheads exported in 1790, Virginia was averaging less than 25,000 in the decade before 1860,[3] yet she was prospering as she had never prospered before. To be sure a local market for the staple had developed in the meantime—a good deal of tobacco was being

manufactured in Virginia—but it had largely passed from the specialized staple of a plantation to one of the crops of a diversified farm.

When Beverly wrote, he declared that the best tobacco land in Tidewater Virginia had been taken up.[4] By 1745 tobacco planting was important seventy miles above the fall line of the Appomattox River, and the Piedmont country rapidly became the chief area of Virginia's tobacco culture. In the first half of the next century cultivation became centered in the south central part of the state around Petersburg.

The movement into the Piedmont came as the soil of the tidewater became exhausted and planters who had not secured good reserve land near their ancestral estates had to move inland. The virgin soils of the Piedmont were not, however, as excellent as those of the tidewater and, in addition, transportation costs were higher. The result was more complaint against the exactions of British merchants and against the mercantilistic policy of Great Britain. Political revolution was accompanied by an agrarian revolution which finally substituted grains and a diversified agriculture for tobacco. This change began first in the tidewater, in the wake of the newer plantation frontier in the Piedmont.

The beginnings of these changes appeared long before the Revolution. About 1741 a few planters turned to indigo, flax, hemp, and wheat. But none of these became important except wheat, which in 1745 was replacing tobacco in some sections and was being exported in the decade before the Revolution in amounts averaging between 40,000 and 80,000 bushels annually. In 1770 Roger Atkinson called it a "kind of second staple" and predicted that it would soon exceed tobacco in importance.[5] But the loss of the wheat market in the West Indies as a result of the Revolution turned planters back to tobacco. Until about 1840 they alternated between tobacco and wheat as European conditions raised or depressed the price of now one and then the other.

In addition to market variations probably the most important factor in Virginia's agricultural changes was soil exhaustion. The "lusty soyle" of the Tidewater, as John smith called it, had produced the large yields of excellent tobacco which had made early Virginia prosperous. But tobacco is a plant peculiarly hard on the soil, and the early planters knew little of the methods of good husbandry and apparently cared even less. With plenty of unappropriated fertile land available it was easier and less expensive to abandon old lands and take up new ones than to restore worn-out soil. Land was regarded virtually as a part of the current expenses rather than as capital.

The tobacco plant makes heavy inroads upon the nitrogen and potash resources of the soil, and it was usual to remove the entire growth from the field. These factors combined served to cause a rapid decline in the available food materials for tobacco and other plants. Since tobacco employed all the labor force and monopolized the best lands, there was little possibility that livestock, through a supply of animal manure, might serve to renew and maintain soil vigor. A field was unusually resistant if it withstood more than three years of such devastating cultivation. Moreover, as new fields were cleared for additional crops or tobacco, the old fields were left neglected, without any covering or vegetation to prevent washing, with the result that they were quickly cut into deep gullies, or further depleted by a succession of corn crops. Such a process of exploitation as this could not go on forever and must have brought loss in its ultimate effect.[6]

Foreign visitors to Virginia before and after the Revolution did not fail to comment upon the "cut down, wear out, and walk off" methods of agriculture which they saw. And a native writer lamented:

> Our ridges have become so barren that they do not afford cover for the partridges, and they have followed the soil down the branches and creeks, hovering in the flats. . . . [Virginia's] forests have been swept away, and her great men of genius and worth, together with the hard cultivators of the soil, the bone and sinew of the land, have, by thousands and tens of thousands been driven out of the state, in search of better lands.[7]

There were scores of similar opinions and observations by contemporary writers.[8]

One consequence of the passing of free and fertile land and of soil exhaustion in Virginia, as the writer quoted above says, was emigration. In the first decade of the eighteenth century the movement into the border colony of North Carolina began to be so heavy that the Board of Trade in England asked the Virginia Council for the facts. The Council replied that not only [are] "servants just free go to North Carolina but old planters whose farms are worn out."[9] Emigration from the state did not, however, begin to assume serious proportions until after the Revolution and continued to be serious until about 1840. It was especially heavy in the two decades between 1820 and 1840, when the Tidewater suffered an actual decline in population; only the growth in the western counties saved the state itself from loss.

Most of the migration from Virginia consisted of small farmers without slaves. But the agrarian crisis of low prices and failing lands likewise affected the planters with slaves. The increasing demand for slaves from the cotton frontier was so increasing their value as to make their labor in Virginia more and more uneconomic. The planter found himself in a position where he had little real capital except his slaves. As things were in Virginia he was getting very little return on this capital. There seemed three possibilities open to him. The first was to migrate with his family and slaves to the Southwest, and many made this decision. Or they might meet the market demand for slaves and dispose of their own by sale. A great many planters followed this course and a steady stream of Negroes began to move South with the slave traders. It was estimated in 1836 that during the twelve month period preceding, 40,000 slaves were exported at an average of $600 each, or $24,000,000 for the total number.[10] Finally, the planter had the alternative of staying home, building up his lands, and seeking new crops and new markets. Those Virginians who followed this course finally achieved the agricultural restoration of the state.

Some of the Virginia leaders who contributed to the political revolution against England also contributed to this agrarian revolution. When Washington returned from the French and Indian wars he found his fields exhausted. He determined upon new methods, and was all the more confirmed in this course by the prevailing low prices of tobacco and the high price of the goods which he had to buy from his English factors. In 1768 he wrote his London factor that he had discontinued "the growth of tobacco—except at a plantation or two upon York River" and would in the future "make no more of that article than barely serves to furnish me with goods."[11] Washington turned to wheat as his chief staple and to the improvement of his estate. A letter of instructions to his superintendent at Mount Vernon in 1794 will indicate the lines along which some planters of his day were seeking a way out of agricultural depression and exhausted lands. The superintendent was instructed to give his best efforts to recovering the land from the gullied and exhausted conditions into which it had fallen; to improve and increase the stock to the full extent of the pasturage; to put all low and swampy lands in grass; and to sow enough clover to support working horses and cattle.[12]

After Washington and Jefferson, John Taylor, J. M. Garnett, Edmund Ruffin, and many others continued until the Civil War to experiment and to advocate improvement in agricultural methods. Perhaps the greatest of Virginia's agricultural reformers was Edmund Ruffin, who demonstrated the value of marl as a soil restorer and led the crusade for manuring, rotating, diversifying, and more intensive farming in general. But the new methods were slow in becoming common knowledge.

Between the Revolution and 1820, Virginia's agriculture, in spite of significant changes, continued in the use of colonial crops, markets, and methods. But after 1820 this was ended. "The hold of tobacco was now broken," says Craven, "and complete dependence upon European markets and marketing systems had largely ceased. . . . The large estate had shown its inability to meet the needs of the future and was being, and was still more to be divided."[13]

The period, 1820 and 1840, was one of manifest transition. The exodus from the state was at its height. Agricultural decline was still apparent but the forces of reorganization were centering around the new knowledge of improved methods and results were beginning to show about 1840. From 1838 to 1850 the value of land in Tidewater Virginia increased by $17,000,000, largely due to the new agriculture. Old gullied fields were now growing clover, once abandoned lands were making abundant fields of wheat and corn, and even emigrants were returning.[14] Everywhere intensive agriculture with deep plowing and crop rotation was replacing the older extensive agriculture. The holding of slaves was becoming more diffused with more personal supervision of their labor. Out of a population of 1,047,200 in 1860 there were 52,126 slave-holders. Of these, 114 owned one hundred or more slaves, one-half owned four or fewer, and one-third owned only one or two. There were 250,000 white laborers in the state, a larger proportion than in Ohio.[15]

The decline of wheat after 1840 helped further to diversify agriculture. Diversifying with tobacco, corn, wheat, stock, orchards, and gardens, the small farm was able to compete effectively with the large estate whose early economic advantage had largely passed away. Around the cities the market gardener and the dairy farmer were beginning to get differentiated. It was, indeed, a new life, as Craven says, that had come to Virginia.

> The new life that had come to Virginia and Maryland was more like that in the northern states than like that in the lower South. Even the governor of Virginia advertised her agriculture as no longer "the large plantation systems" but now one of "small horticultural and arboricultural farming." Emigrants from northern states began to move into Virginia and Maryland as agriculture became like that of their own states. . . . Slavery had lessened in importance with the coming of diversified farming, yet on the larger estates it had fitted well into the system and new prosperity gave the owners a feeling that "emancipation (was) an idle dream," and that its accomplishment would overthrow "not only all the material interests of the South, but also the great fabric of modern civilization." . . . The agricultural interests of the state were therefore divided and this agricultural division played its part in the struggle over secession and unionism in this section. As might be expected this line or cleavage ran rather closely along the line of interests based upon the new agricultural life—its markets, its labor, and its social relations.

It is thus apparent that by 1860 soil exhaustion had ceased to be a problem in Virginia and Maryland. The two states had come largely to the small farm and the small farmer. Even where the larger estate still persisted, extensive crops and methods had given way to intensive cultivation or diversified production. New markets had been found and a new prosperity established upon a restored fertility. The War Between the States found the section far removed from the days when an exhausted soil limited her agriculture and sent her sturdy sons fleeing into the West.[16]

The Plantation on the New Southern Frontier

The plantation arose in Virginia as a means of establishing order and of maintaining economic purposes. It arose *in* settlement and achieved stabilization when it became a part of the *mores* and produced attitudes on the part of both planter and slave whereby it was taken for granted as naturally as a family is taken for granted. As the frontier moved west and as river transportation in the interior was utilized or improved, the conditions under which the plantation originally appeared were continually reappearing. But now it was possible to transplant a political institution already full-grown rather than await the evolution of one. The plantation became an institution *for* settlement under conditions for which its fitness had already been tested.

The plantation was accustomed to move. The exhaustion of the soil in Tidewater Virginia forced many planters to seek the fresher lands of the Piedmont. Within the colony of Virginia the plantation sought and found its best opportunities along the advancing frontier, provided transportation for tobacco was possible. But it was the growing demand for a new staple, cotton, and the invention of the cotton gin, that started the plantation on a career of conquest which took it, by 1860, to Texas. It is not asserted that this Southwest movement was an expansion of the Virginia plantation system, for it was not, although Virginia contributed largely to it. But this greater expansion illustrates on a larger scale the nature of the changes that went on within Virginia itself as the plantation system moved into the Piedmont while it underwent modification in the Tidewater. In accounting for the part which the plantation was to play along the cotton frontier it is necessary to understand the nature of the changes that resulted in building up a market for cotton.

At the close of the eighteenth century, cotton came to England from the Levant, India, Brazil, and the West Indies. By this time it had supplanted wool as the chief textile of commerce. "The possibility of an indefinite increase in the raw material supply made cotton especially suitable for machinery which needed large quantities of raw material so that the machines should be constantly employed."[17] Simple inventions which facilitated the spinning of wool and flax were applied to cotton processing before 1741. Since with cotton technical difficulties were more easily overcome, new inventions in its manufacture followed one another in rapid succession—but in spite of these inventions more cotton came in than could be handled.

The inventions are directly related to the pressure of supplies on manufacture in England. In 1738, for example, when about 1,500 thousand pounds of cotton were imported into England, Kay invented his "drop-box," Hargreave his "spinning jenny," and calico printing was introduced into Lancashire. Imports rose to about 5,000 thousand pounds in 1775, and Crompton's "mule" was invented a few years after. In 1783 imports totaled 9,735 thousand pounds, and cylinder printing appeared the same year. In 1785, 18,400 thousand pounds of cotton entered, and the steam engine was applied to textile manufacture, Cartwright invented his power-loom, and oxymuriatic acid was used in bleaching cotton goods.[18]

By the close of the century these inventions made it possible to consume all the raw material imported into England, and the demand began to exceed the available supply. Pressure was now applied to the frontiers of production for a larger supply.[19]

At this point was made the only important invention in the cotton textile industry to be furnished by an area where the raw material was produced. In Georgia, in 1793, Eli Whitney invented the cotton gin, and the South, which in 1791 produced only about two million pounds out of the world's total 490 million,[20] immediately jumped forward. Exports increased from only 348 bags in 1794 to over 2,000 bags the following year, an increase of over 500 per cent.[21] With the exception of improvements on the cotton gin no further inventions were made or immediately needed in the production of raw cotton. "Except during the interruptions occasioned by the embargo of 1808, the war of 1812–14, and the small crop of 1823–24, the supply of American cotton down to 1826 kept rather ahead of consumption."[22]

Now pressure was again upon the industrial center in England, and more inventions to utilize the new supply were necessary. In 1796 with imports at 22 million pounds, Miller made important improvements in the power-loom. In 1797 imports amounted to well over 23 million pounds, and a scutching machine appeared. As imports grew in volume still more inventions were made to handle the flood of cotton.[23]

On the Southern frontier occurred changes, without mechanical inventions, that answered to the industrial changes taking place in England. Power farming, i.e., the plantation with slave labor, already had been applied to the cultivation of cotton, and now with each mechanical improvement in England and with each new increase in demand for the staple, the plantation settled and opened up new producing areas. Behind the rapid territorial expansion into the Southwest was the cotton plantation, large or small, pushing it on. The plantation settlement of the South, therefore, was not the mere working of a restless instinct, but a direct effect of an expanding market for a staple which the world needed.

After the invention of the cotton gin, cotton production expanded into the Piedmont region of South Carolina and east central Georgia. Before 1815 planters had pushed into Tennessee and begun growing cotton there. In 1804 the purchase of the Louisiana territory opened up a new area for cultivation, and by 1819 the fertile lands of Louisiana, Mississippi, and Alabama were contributing large quantities of cotton. Between 1830 and 1840 production began to center in the fertile area along the lower Mississippi River and the Black Belt of Alabama. In the next decade Texas and Arkansas turned to cotton.

The spread of cotton production in the South, however, is not a good index of the expansion of the plantation, since many small farmers produced the staple. A better index is the distribution of Negro slaves from decade to decade. For the presence of the Negro slave signified the presence of capital as well as labor, and of the institutional attitude which, subjectively, was the plantation. From the Atlantic tidewater there was, after 1810, a jump in the distribution of slaves over the Sand Hills and Pine Barrens of South Carolina and Georgia into the Piedmont region of those states. Slavery was established along the southern lowlands of the Mississippi River at an early date since they are adjacent to the coast and because of the scheme of French commercial companies. In the Central Cotton Belt, a large belt running south of the mountains from southern Tennessee and northeastern Mississippi through central Alabama, arose the densest slave concentration. By 1860 slavery had spread into eastern Texas and Arkansas.[24]

The distribution of slaves along the cotton frontier was accomplished in two ways, by the migration of planters with their *familia,* and by the domestic slave trade. As to the first, there is a significant story told by Isaiah Montgomery, a Negro leader in Mississippi after emancipation. Montgomery, while urging the building of the Negro town of Mound Bayou, reminded his followers that they had formerly made settlements for their white masters and now they could just as easily make a settlement for themselves. To illustrate his point he told the story of a planter who brought his slaves to Mississippi, left them there to clear the plantation and grow a crop of cotton, while he returned to the old home on other business.[25] Nothing could better demonstrate the institutional character of plantation settlement which simply carried an older feudal order into a new country.

Table II

Textile Inventions and the Cotton Trade

Inventions, etc[1]	Year	Year	Imports into United Kingdom[1] (000 omitted) POUNDS	Production of United States[2] (000 omitted) POUNDS	United States[2] Exports (000 omitted) POUNDS	Exports from U.S. to United Kingdom (000 omitted) POUNDS
Kay's "Fly-shuttle."	1738	1701–05	1,170*			
Paul's "spinning by rollers."		1716–20	2,173*			
Paul's improved carding machine	1748	1741	1,645			
Kay's "Drop-box."	1760	1751	2,976			
Hargreave's "Spinning Jenny." Calico printing introduced into Lancashire.		1764	3,870			
Arkwright's first patent. Watt's first patent.	1769					
Hargreave's first patent.	1770					
Arkwright's mill built at Crompton.	1771					
Arkwright's and Wood's carding machines.	1773					

Arkwright's second patent.	1775	1771–75	4,764	
Crompton invents mule.	1779			
Watt's second patent.	1781	1776–80	6,766	
Muslins first made.		1781	5,196	
First import of Brazil cotton—very dirty.				
Watt's further improvements.				
Export of cotton machinery prohibited: penalty £500.				
Cylinder printing invented.	1783	1782	11,828	
Quantity of cotton imported	1784	1783	9,735	
from United States seized on ground that it was not American produce.		1784	11,482	1 1
Arkwright's patents thrown open. Steam-engine first applied to cotton factory. Cartwright invents power-loom. Oxymuriatic acid first applied to bleaching cotton goods.	1785	1785	18,400	2 2
Arkwright knighted.	1786	1786	19,475	1 1
Cartwright's improved power-loom	1787	1787	23,250	16 16

Table II *continued*

Inventions, etc.[1]	Year	Year	Imports into United Kingdom[1] (000 omitted) POUNDS	Production of United States[2] (000 omitted) POUNDS	United States[2] Exports (000 omitted) POUNDS	Exports from U.S. to United Kingdom (000 omitted) POUNDS
East India Company pressed to push growth of cotton in India.	1788	1788	20,467		58	58
		1789	32,576	1,000	126	126
Arkwright adopts steam in his factory.	1790	1790	31,477	1,500	12	12
		1791	28,706	2,000	189	189
Improvement in mule by Kelly.	1792	1792	34,907	3,000	138	138
Whitney invents saw-gin. Improvement in mule by Kennedy.	1793	1793	19,040	5,000	488	487
		1794	24,358	8,000	1,602	1,602
Improved saw-gin by Whitney.	1795	1795	26,401	8,000	6,276	6,276
Improvement in loom by Miller.	1796	1796	22,126	10,000	6,107	6,106
Scutching machine invented.	1797	1797	23,354	11,000	3,788	3,788
Tennant's patent for bleaching.	1798	1798	31,880	15,000	9,360	9,360
		1799	43,379	20,000	9,532	9,532

	Year				
	1800	60,345	35,000	17,790	16,180
	1811	90,000	80,000	62,186	38,073
Horrock's dressing machine. 1813	1813		75,000	19,400	9,279
	1821	165,000	180,000	124,893	83,419
	1831	235,000	354,547	270,980	210,875
Robert's self-acting mule. 1832	1832		355,492	322,215	229,733
Bullough's improved power-loom. 1841	1841	500,000	644,172	517,628	338,344
Howe's sewing machine. 1846	1846		863,321	685,042	453,074
	1851	780,000	1,021,048	827,303	789,998
	1861	1,275,000	1,934,546	1,531,850	1,037,582

1. Thomas Ellison, *The Cotton Trade of Great Britain*, London, 1886, p. 29; Harold Rugg, *Changing Civilizations in the Modern World*, Boston, 1930, p. 62.

2. M. B. Hammond, *The Cotton Industry. An Essay in American Economic History.* (American Economic Association). Appendix I.

*L. C. A. Knowles, *The Industrial and Commercial Revolutions in Great Britain during the Nineteenth Century*, London, 1926, p. 47. Quoting from Guest, *Compendious History of the Cotton Manufacture*, 1823, p. 51.

But these later Southern frontiers differed from the early Virginia frontier in certain important respects.

Between the white settlers on the further frontiers competition was free and democratic. They became adjusted to each other by establishing themselves on their individuals farms, with each settler almost completely sovereign over his own domain. He did not expect to be interfered with or to interfere in the affairs of others. With racially and culturally homogeneous neighbors a minimum of formal government was needed, for the *mores* constituted sufficient control of relations. But when into this situation came many who were racially different, which itself offended the prejudices of the democratic white farmers, and who, moreover, were regarded as dangerous, or potentially dangerous, but who were not inclined to assert themselves, the democracy of the frontier broke down. Such individuals could not assume the role of free competitors but could be accommodated only through subordination to other individuals in good standing who would be responsible for them.

In Virginia this eventuated in Negro slavery when coupled with the planter's need for labor. But on the newer frontiers the Negro arrived, in the first place, as a completely subordinated slave, not as an indentured servant who in a few years might be free. Thus the period of determining the status or the alien element and of eliminating it as a contender for land was absent on the newer frontiers. Likewise the period which in Virginia had witnessed the evolution of a planter aristocracy while Negroes were becoming slaves was absent on the newer frontiers. Such an aristocracy was transplanted to these frontiers, and its appearance aroused the prejudices of the democratic white farmers who had previously settled there.[26] This occurred when the plantation institution, with both its Negro peasantry and its aristocracy, was moved to a new frontier with its old moral and social order remaining intact.

The second way in which slaves were distributed along the cotton frontier was by means of the domestic slave trade. This necessarily involved individuation until incorporation into a new plantation society took place. With fertile lands virtually free, or at least very cheap, a planter who wished to profit on a large scale would desire more slaves than had migrated with him, and he was willing to pay high prices for them, higher prices than slaves could command in the soil-exhausted areas of Virginia and other eastern communities. Of the Southwest Ingraham wrote: "Cotton and negroes are the constant theme—the ever harped upon, never worn out subject of conversation among all classes."[27] On the frontier of the Southwest, according to Bancroft, traders might expect considerably more than $2,000 for slaves for whom they paid $1,600 or $1,640 in the old states.[28]

While the plantation tended to assume a domestic and patriarchal character in the older communities, on the Southwest it was developing into a flourishing institution. Census figures show the rapid increase of the slave population in the newer communities.[29] Between 1850 and 1860 Texas, the last slave state, increased in total population 184 per cent while her slave population increased 215 per cent. In Virginia, on the other hand, from 50 per cent of the total population in 1782, the Negro population declined to 43 percent in 1830 and to 37 per cent in 1860.[30]

The plantation was at its height where there was still free or available land with opportunities for transporting and marketing the staple. This condition was to be found only on the frontier. When all the available land was occupied free labor could begin to compete with some degree of success against slave labor and, says Fleming,

> a slow decline [of the plantation] was later sure to follow. The planter freed slaves or abandoned a district when economic conditions forced him to do so; then slowly came in free labor upon small farms. This slow movement was taking place in the South Atlantic states, especially in Maryland, Virginia, and North Carolina. In the older states large holdings of slaves tended to break up into smaller holdings, and owners worked with their men. Slave labor was then used on farms rather than on plantations, and the latter tended to decrease in relative importance.[31]

In the absence of exact statistical information of farm units perhaps the best evidence for the partial decline of the plantation in the wake of frontier settlement is given in the census data on the growth of the free colored population by states. Since it is highly improbable that many free Negroes, if any, migrated to the frontier from older plantation communities, it follows that their appearance in a plantation area was a sure sign of the beginning of the institution's disintegration, of the passing of free land.[32] Both in the border states and in the cotton states of the Atlantic seaboard free Negroes were present in large numbers, but in the newer cotton states they were insignificant numerically. In Maryland, between 1850 and 1860, while the free Negro population was increasing by 12 per cent, the slave population was actually declining.[33] Especially in the border states, and to some extent in the South Atlantic states, plantations in 1860 were being divided into farms, new villages were appearing, and small industries were being organized.[34] It was in Mississippi, Texas, and Arkansas that the plantation in 1860 was at its height.

The changes in plantation economy which were under way in 1860 were not totally transformed by the Civil War and emancipation. The destruction of capital and property in the form of slaves, together with the right of free movement which the Negro gained, retarded plantation progress along the Southwest frontier and hastened the trend toward peasant proprietorship in the older South. The

Table III

Population Changes in the South

1790–1860

Date	1790	1800 Pop.	1800 % Inc.	1810 Pop.	1810 % Inc.	1820 Pop.	1820 % Inc.	1830 Pop.	1830 % Inc.	1840 Pop.	1840 % Inc.	1850 Pop.	1850 % Inc.	1860 Pop.	1860 % Inc.
Virginia															
White	391,524	443,356	13.24	458,159	3.33	482,849	5.38	537,216	11.25	537,952	.13	616,069	14.52	691,773	12.28
Slave	287,959	338,624	17.59	381,680	12.71	410,029	7.42	452,084	10.25	430,499	4.77*	452,028	5.00	472,494	4.52
Free-colored	12,254	19,598	59.93	29,292	44.36	35,470	21.09	45,181	27.37	46,809	3.63	51,251	9.48	55,269	7.03
Maryland															
White	208,649	216,326	3.67	235,117	8.67	260,233	10.67	291,108	11.86	318,204	9.30	417,943	31.34	515,918	23.44
Slave	103,036	105,635	2.52	111,502	5.55	107,397	3.68*	102,994	4.09*	89,737	12.87*	90,368	.70	87,189	3.51*
Free-colored	8,043	19,587	143.52	33,927	73.21	39,734	17.14	52,938	33.24	62,078	17.26	74,723	20.36	83,942	12.33
South Carolina															
White	140,178	196,255	40.03	214,196	9.14	237,440	10.85	257,863	8.60	259,084	.47	274,563	6.36	291,300	6.09
Slave	107,094	146,151	35.53	196,365	34.35	258,475	31.62	315,401	22.02	327,038	3.68	384,984	17.71	402,406	4.52
Free-colored	1,801	3,185	76.84	4,554	42.98	6,826	49.89	7,921	16.04	8,276	4.48	8,960	8.26	9,914	10.64
North Carolina															
White	288,204	337,764	17.19	376,410	11.44	419,200	11.36	472,843	12.79	484,870	2.54	553,028	14.05	629,942	13.90
Slave	100,572	133,296	32.53	168,824	26.65	204,917	21.37	245,601	19.85	245,817	.08	288,548	17.38	331,059	14.73
Free-colored	4,975	7,043	41.56	10,266	45.76	14,712	43.30	19,543	32.83	22,732	16.31	27,463	20.81	30,463	10.92
Georgia															
White	52,886	102,261	93.36	145,414	42.18	189,566	30.36	296,806	56.57	407,695	37.36	521,572	27.93	591,550	13.41
Slave	29,264	59,406	103.00	105,218	77.11	149,656	42.23	217,531	45.35	280,944	29.15	381,682	35.86	462,198	21.09
Free-colored	398	1,019	156.03	1,801	76.74	1,763	2.10*	2,486	41.00	2,753	10.74	2,931	6.46	3,500	18.41

Kentucky															
White	61,133	179,873	194.23	324,237	80.25	434,644	34.05	517,787	19.12	590,253	13.99	761,413	28.99	919,484	20.76
Slave	12,430	40,343	224.56	80,561	99.69	126,732	57.31	165,213	30.36	182,258	10.31	210,981	15.15	225,483	6.87
Free-colored	114	739	548.24	1,713	131.79	2,759	61.06	4,917	78.21	7,317	48.81	10,011	36.81	10,684	6.72
Tennessee															
White	31,913	91,709	187.37	215,875	135.39	339,927	57.46	535,746	57.65	640,627	19.57	756,836	18.13	826,722	9.23
Slave	3,417	13,584	297.54	44,535	227.84	80,107	79.87	141,603	76.76	183,059	29.27	239,459	25.34	275,719	15.14
Free-colored	361	309	14.40*	1,317	326.21	2,737	107.82	4,555	66.41	5,524	21.27	6,422	16.07	7,300	13.67
Mississippi															
White	....	4,446	....	16,602	273.41	42,176	154.04	70,443	67.02	179,074	122.52	295,718	65.13	353,899	19.60
Slave	....	2,995	....	14,523	384.90	32,814	125.94	65,659	100.09	195,211	197.31	309,878	58.74	436,631	40.90
Free-colored	....	159	....	181	13.83	458	153.03	519	13.31	1,366	163.19	930	31.91*	773	16.88*
Louisiana															
White	....	....	....	34,311	....	73,383	113.87	89,231	21.59	158,457	77.58	255,491	61.23	357,456	39.90
Slave	....	....	....	34,660	....	69,064	99.26	109,588	58.67	168,452	53.71	244,809	45.32	331,726	35.50
Free-colored	....	....	....	7,585	....	10,476	38.11	16,710	59.50	25,502	52.61	17,462	31.52*	18,647	6.78
Alabama															
White	....	....	....	....	....	85,451	....	190,406	122.82	335,185	76.03	426,514	27.21	526,271	23.38
Slave	....	....	....	....	....	41,879	....	117,549	180.68	253,532	115.68	342,844	35.22	435,080	26.90
Free-colored	....	....	....	....	....	571	....	1,572	175.30	2,039	29.70	2,259	11.83	2,690	19.07
Arkansas															
White	....	....	....	....	....	12,579	....	25,671	104.07	77,174	200.62	162,189	101.60	324,143	99.23
Slave	....	....	....	....	....	1,617	....	4,576	182.37	19,935	335.64	47,100	136.26	111,115	135.91
Free-colored	....	....	....	....	....	59	....	141	138.98	465	229.78	608	30.75	144	76.31*

*Loss

trend in Texas since the [Civil] War has been in the direction of plantation extension and growth with white, Negro, and, more recently, Mexican tenants and
laborers.

Cropper tenantry, the post-war substitute for slavery, is a form of farm tenantry in America known only to the South. It involves the direct supervision of
the planter or overseer. The substitution of cropper tenantry for slavery in plantation organization has brought the poor white back on the estate as a laborer and
to that extent has served to put him on the same level as the Negro. In addition,
the change has organized plantation production by families, each with a one- or
two-horse farm, or subdivision of the plantation, and some degree of responsibility for it during the period of tenantry.[35]

Many tenants, white and black, have been able to purchase and become
owners of these small farms, while others have gained proprietorship without
ownership. It is possible that, at least in the South Atlantic states, the degree of
independent, or near-independent, proprietorship, even within the forms of the
old plantation organization, is greater than is usually assumed. Unless new inventions and new methods of operation are introduced on a large scale, it is probable that this present trend toward peasant proprietorship will continue.[36]

With the partial disintegration of the plantation since the Civil War has
[come] a corresponding increase in towns and villages. In the old plantation areas
these have grown up around country stores. They indicate the development of
local divisions of labor which extend from the towns out into the countryside.
These town and village focal centers have helped further to break down the plantation social world. The redistribution of population into towns and villages has
produced numerous social movements, educational, agrarian, and political. These
seem to date from about 1880, when the new order was well enough established
and far enough away to stimulate memories of the old plantation order, the age
that was golden. In romantic retrospect the antebellum plantation was pictured in
terms of manorial grandeur, genteel manners, fair ladies, gallant gentlemen, and
devoted serfs.[37] Also in each Southern state demagogue politicians with new issues
began to appear with appeals to a new constituency, the white democracy of the
towns and farms.

Not only emancipation but the rise of towns tended to tear apart the web of
old plantation relations. There came a new competition between white and Negro
workers on the plantations and in the towns. There were new conflicts and prejudices arising out of this competition. The new adjustment was sought by the
demagogue (leaders of the whites in terms of the old pattern of plantation) relations, the subordination of one race and the superordination of the other. Habit

and custom and dominant interests were with them, and they have largely succeeded. Habituated to a menial position in slavery, the mass of Negroes have easily accepted subordination in the new order. But there were those who did not
accept it because they could not. For the most part these have been mulattoes.
These mulatto leaders, with "heart black and mind white," denied parity with the
white man, have identified themselves with the cause of the mass of Negroes, but,
unlike the mass, they have not been content, cannot be content, with a subordinate status.[38] For them social relations must be redefined on a plane where the
"differently born" can find security and recognition. Until then they must remain
native aliens. The greatest of these leaders, Booker T. Washington, born a slave on
a Virginia plantation, devoted his life within the South as head of Tuskegee
Institute for Negroes to the problem of redefining white and Negro race relations
in accordance with his formula: "In all things that are purely social we can be as
separate as the fingers, yet one as the hand in all things essential to mutual
progress."[39]

In spite of the fact that economic and social changes in the plantation South
have been great, they must not be exaggerated. Underneath all, even the divergent
tendencies, the old plantation pattern continues on in the New South. Rule by
the same majority group occurs only in the South, elsewhere the membership of
the "majority" changes from one issue to another. This is because the issues that
appear as problems elsewhere in the United States are not problems in the agrarian and feudal order of the section where they are settled in the *mores*. Relatively
few flounder puzzled and perplexed, trying to find new values which give meaning to life because the plantation already has defined those values and whether
they are ethically high or low they are accepted by the folk because they work.
This is the cultural order of the South, the heritage of its plantation civilization.

6

The Natural History of the Plantation

Geographical Isolation and Culture

The history of the plantation in Virginia and in the South, because of its wealth of human interest, is interesting and instructive for its own sake alone. It aids in securing an understanding of the Southern experience, and of the social attitudes and cultural heritage which have grown out of that experience. Nevertheless, a study of a single community or epoch is of limited scientific value unless it is so formulated as to be readily compared with similar situations in other times and places. Such a formulation gives the study a general, rather than a special, significance.

As distinct from the history of a particular country, or of a particular period of time, the history of an institution raises general questions and challenges comparisons with corresponding institutions elsewhere. A result of such questions and comparisons is almost invariably some attempt to strip away what appears to be fortuitous and non-essential and to search for what is typical and general. This, at least, is what happened when the family, the state, the rural community, the city and other institutions came to be studied comparatively. The institution which seems to emerge from a comparative study is not, however, a static thing; it is not a dead museum specimen. There also emerge similarities in lines of development, for the typical institution has not only a history, but also a natural history.[1]

The plantation has been regarded in this study as a social institution. In one historical instance, Virginia, we have studied the facts concerning its origin and development. What conditions explaining the plantation in Virginia are characteristic of plantations in general and what conditions are peculiar to it? How far, in other words, is the Virginia plantation typical?

So far as the writer is aware, no strictly comparable study of the plantation in an area other than Virginia has been made. Nevertheless, a general knowledge of the history of other plantation areas suggests what is typical about the institution in the one area selected for more intensive study. Isaiah Bowman indicates that situations similar to those in which the Virginia plantation arose and established itself in the seventeenth century are continually reappearing on other frontiers.

The agricultural frontier of Brazil has been carried far into the wilderness by Negroes and Indians, and by mestizos as well. Freeman and fugitives (both criminal and political) still form, as they have always formed, a thin fringe of populated land "beyond the habitual reach of the law" in the back country of every agricultural community. Though calling themselves settlers they are really squatters with shifting agriculture as their mode of life. The shift takes place naturally or through the compulsion of the man who has title to the land and who only awaits the creation of capital values to assert his ultimate rights. Some of the more restless settlers move on to new locations, others accept more or less willingly a feudal relation set up by the owner. The law of Brazil protects the squatter in the enjoyment of the buildings and crops that he has created, and he can no longer be driven from the land without remuneration. The frontier plantation, remote and feudal, can also be a means of protection to a serf. The lawless owner may himself represent the law to the ruder societies or individuals about him.[2]

The differences between the plantation frontier in Brazil and other plantation frontiers, past and present, are accounted for by differences in ecological position, geography and natural resources, and the particular traditions and purposes of the people who come to occupy the land. "The background of the pioneer is an important part of his equipment," continues Bowman in accounting for the peculiarities of the plantation on the Brazilian frontier:

From what social and economic systems came the South American who now lives on the fringe of settlement? I have found throughout South America the definite assumption that aristocracy (and in most countries the church also) is a rational if not a necessary creation. In precisely its historical forms the aristocratic point of view was advocated in Spain down to the revolution of 1931. Church and State for centuries consolidated their power in the long struggle against the Moor. Soon after the closing of the chapter the exploitation of the New World came into full swing. The knightly cavalier no less than the soldier of fortune of lowly birth found here the means of realizing traditional cultural ideals which included, on the emotional side, religion and war. It happened that where there was gold and silver there was also an organized agricultural folk that was tied to the soil and had no place or refuge, no forests in which to hide, no untamed open plains. The conquerors became the overlords, that is, the aristocrats of New World society. They fastened themselves upon the native system, adapted or overturned it, and, working together closely, priest and cavalier became the cornerstones of the new order.[3]

The particular tradition and purposes of those who seek to bring the land of the frontier into higher uses thus partly account for the different between plantation areas. It is necessary, therefore, in the history of particular plantation areas to take these traditions and purposes into account. The history of the plantation as an institution begins with the migration and contact of peoples of diverse races and cultures. These races and cultures may be considered as products of isolation.

> Race relations are, or were, primarily geographic rather than human and social. The races grew up in isolation and acquired distinct racial characteristics slowly by adaptation and by inbreeding. Man, like every other animal, has been and is a creature of his environment, even when that environment has consisted largely of other men. Biological and inheritable differences represent man's responses to the kind of world in which he has learned to live. They are, so to speak, his biological capital; the accumulations of successive generations given in their struggle to live.[4]

Not only the biological earmarks of the people we call "races" result from geographical isolation, but whatever is distinctive in the various group cultures likewise develops in the experiences of isolated groups. What is most distinctive and peculiar about races and peoples, what a people take for granted, their culture, in other words, is an effect of the particular conditions under which they have lived with physical nature and with each other. Their architecture, art, religion, dress, language, and food habits are all part of their cultural life, of that with which they are, paradoxically, familiarly unfamiliar.

In isolation the customs and traditions of a people plead for themselves, and there is little or no change in them. But when diverse racial and cultural groups come together the effect is, in some greater or less degree, to break down the established cultures and lay the basis for a new one. Men of diverse race and creed are united for the immediate purpose at hand and new groupings arise. The sequence of processes which are set into operation by these racial and cultural contacts has been summarized by Park in the following excerpt:

> The race relations cycle which takes the form, to state it abstractly, of contacts, competition, accommodation and eventual assimilation, is apparently progressive and irreversible. Customs, regulations, immigration restrictions and racial barriers may slacken the tempo of the movement; may perhaps halt it altogether for a time; but cannot change its direction, cannot, at any rate, reverse it.[5]

Where the situation is favorable, the plantation arises during the course of the race relations cycle, as Park has formulated it, as a form of accommodation, that is, as a way of establishing toleration and division of labor between divergent races and cultures. The plantation, however, is more than an accommodation of racial groups *per se*. The organization is accomplished around a territorial division of labor in the world community. It is an economic as well as a social institution and as such is involved in a cycle of change involving and accompanying changes in natural resources, economic costs of production, labor supply, and market relations. This cycle has been variously described as the passing of the frontier, the transition from plantation to peasant proprietorship, or the transition from open to closed resources. The race relations cycle and the economic cycle seem to parallel each other, at least conceptually, and to interact with each other at every stage. They are, respectively, the sociological and the ecological aspects of a single development. However, we shall endeavor, in outlining the natural history of the plantation, to keep both aspects before us.

Ecological Changes and Race Relations

An effect of international commerce has been to knit the various areas and peoples of the world into an organic world community. Metropolitan markets arose as functional centers of a spider-web pattern by means of which the world community is integrated.[6] Relations between the parts are maintained through communication and transportation. The tendency toward an economic equilibrium is continually being affected by changes in means and methods of transportation and other factors which enter into the competitive relations between areas.[7] Ecologically, the world is shrinking, as McKenzie puts it, but not evenly. "Some places are coming closer together" while "others are remaining stationary or even becoming more remote."[8]

The agricultural economy of an area in the world community is not merely a response to immediate factors of soil, climate, topography, etc. Even more important than these, in many cases, is the market relation; a particular economy cannot continue to exist which is not adjusted to this relation. Change in ecological distance or position, therefore, brings with it an inevitable change in the form of economy.[9] This may involve, as is usually the case where colonization takes place, a displacement of a self-sufficient native economy by an economy based upon the metropolitan market, or it may mean the substitution of one commercial agricultural economy for another.

In the modern world community the resources of an area may be described as *open* or *closed*, depending upon the competitive advantages or disadvantages of the area over other areas in marketing them. An index of the extent to which an area is one of open resources may be found in the extent to which the area is attractive to outside capital seeking investment opportunities. Capital flows most freely to those areas promising the greatest returns.[10] Improvements in the means of transportation and communication, or changes in commercial routes, have the effect of closing or opening the resources of areas dependent upon the market relation.

Not only capital but population moves toward the areas of open resources. In the period of slavery and the slave trade a larger part of the capital was invested in the labor to be employed in exploiting resources. In the seventeenth century the American plantation was begun with capital, management and labor originating in one source, England; it later became one point of a triangular relation whose other two points included Europe, which furnished the capital and management, and Africa which furnished purchased bond labor. In both cases capital and labor were involved in one migration and appeared simultaneously on the frontier. At the present time the migration of labor is regulated primarily by differential advantages of areas and industries in wage and salary scales, and its direction is determined by the movement of capital which precedes it.[11] The differentiation in source of capital and of labor, which began with the slave trade, is now the usual pattern, so that the confrontation of capital and labor under [a] plantation economy is commonly on a frontier foreign to both.

Ecological changes are thus invariably accompanied by the migration and redistribution of population. This migration, unlike the tribal wanderings of primitive peoples, is generally individual or "collective."[12] The recruiting of unskilled labor for plantation work is a problem of securing it at minimum wages and not one of securing a homogenous racial or cultural population. The conditions under which such cheap, unskilled labor is recruited are such as to promote racial and cultural diversity.[13] On the plantations this heterogeneous labor supply is brought under the management of men representing still other cultural and racial groups who face the problems of organizing and controlling the labor to serve the purposes for which capital [was] originally invested and for which they are responsible. A secular society with impersonal relations is established and the stage is set for these relations to take the form of an institution. The whole process is an aspect of the general change from an indigenous world of isolated races and cultures to one based upon economic relations with metropolitan markets in which there is a highly organized flow of commodities and peoples. As Benton Backaye expresses it, it is change from a "quiltwork or varied cultures" to a "framework or world-wide standardized civilization."[14]

Adaptation and Accommodation to a New Habitat

The most elementary aspect of race relations is perhaps that of biological competition which means, according to Park, "the struggle to determine, within any geographical area, which race and which races are to survive."[15] This struggle continues until an equilibrium is reached.

The question of racial adjustment to different humidities, temperatures, and degrees of seasonal variety, which are the elements of climate, has led to an almost endless amount of discussion and controversy.[16] Acclimatization involves not merely living away from a homeland, but also competing successfully with natives or members of other races, both biologically and economically.[17] It is adaptation to a different climatic situation, but it is measured largely in terms of success or failure in competition against those with whom new settlers come in contact. Acclimatization is, therefore, a relative matter and is bound up with such factors as the culture and standard of living of the settlers.

The problem of acclimatization is usually discussed[18] in connection with white settlement in equatorial regions. Many writers believe that white acclimatization in the tropics is constitutionally impossible. Keller, as we have already noted, assigns this as the reason for the predominance of the plantation institution in the tropics.[19] It is possible, however, that the difficulties of white settlement in the tropics are largely cultural in nature, and "ability to stand the climate," imputed to another group as a racial characteristic, becomes a part of the ideas on fundamental racial differences and is used to justify the racial division of labor. Southern planters so justified Negro slavery, in part, and the necessity of their summer residence by the sea or in the mountains, but they always left a white overseer on their plantations.

Nevertheless, biological competition is a fact which cannot be completely argued away. Involved in this competition, and, of course, bound up with the question of acclimatization, is that impersonal and for the most part unconscious struggle which depends upon relative immunity to disease. Disease is an index of adaptation and accommodation to the habitat, "the name we give to a group of processes by which the size of a population is adjusted so as to enable it best to utilize the available means of subsistence."[20]

In earlier plantation areas, such as Virginia, little was done to combat the ravages of scurvy, plague, yellow fever, or malaria and these and other diseases took a great toll of life. In the southern part of the United States, Negro slaves apparently were able to withstand them somewhat more successfully than whites.[21] The reverse seems to have been true in the northern colonies.[22] In modern plantation settlements, on the other hand, an effort is made to control or prevent disease through sanitary and medical measures.[23]

Under the conditions in which plantation settlement takes place, the individual migrant, whether planter or laborer, has the character of a stranger in the community. Sombart finds that the settlement of America by men without sentiment for its natural landscape was an important factor in the rise of capitalism. "The assumption . . . forces itself upon us," he says: "that this particular social condition—migration or change of habitat—was responsible for the unfolding of the capitalist spirit." This is due to a "breach with all old ways of life and all old social relationships."

> Indeed, the psychology of the stranger in a new land may easily be explained by reference to this one supreme fact. His clan, his country, his people, his state, no matter how deeply he was rooted in them, have now ceased to be realities for him. His first aim is to make profit. How could it be otherwise? There is nothing else open to him. In the old country he was excluded from playing his part in public life; in the colony of his choice there is no public life to speak of. Neither can he devote himself to a life of comfortable, slothful ease; the new lands have little comfort. Nor is the newcomer moved by sentiment. His environment means nothing to him. At best he regards it as a means to an end—to make a living. All this must surely be of great consequence for the rise of a mental outlook that cares only for gain; and who will deny that the colonial activity generates it?[24]

In this spirit, the migrant stranger exploits the land with a ruthless hand. Says Keller:

> It is not surprising to find the system of plantation-culture a ruthless and wasteful one, not only of soil but of men. It is what the Germans graphically denominate *Raubbau*. Agriculture presents the extensive rather than the intensive form, with all which that implies of non-restoration of the soil, even non-rotation of crops, etc. Frequent and protracted absenteeism of sick or indifferent owners—who are often, indeed, mere shareholders in a company—has played its part in mismanagement and waste. A wide-spread indifference or cynicism respecting the future of the human working-animal has prevailed; he has been regarded in general as an insentient factor in the accumulation of wealth.[25]

Another consequence of the conditions under which labor is recruited for the plantations, and of the spirit of individualism and anarchy which pervades the plantation settlement, is racial miscegenation. Labor is imported as a utility but the creatures thus imported not merely breed but they interbreed with their masters and employers, all the more, of course, when there is a disproportion of the sexes. "Racial hybrids," says Park, "seem to be one of the invariable accompaniments and consequences of human migration. Hybridization is probably, therefore, a mathematical function of the geographical mobility of peoples."[26]

Settlement in the human community, like settlement in the plant community, includes processes that are subsequent to actual residence in the new area. But unlike the plant community, settlement in the human community is not complete with the establishment of a biological equilibrium. As the individual members of different racial and cultural groups seek to act in the new situation, each in terms of his own wishes and interests, they set in operation processes that lead to the establishment of some sort of working relations between them. The plantation thus arises in such a frontier settlement as a means of control. In the absence of cultural homogeneity this control is political, that is, it is based upon the more or less forced subordination of the majority to the few who are strong enough to gain authority and maintain it. When authority is used to apply the energy of the subordinated to the task of continuous and specialized agricultural production for the world market the plantation passes into another stage of its life-history.

Agricultural Specialization and Racial Stratification

The development of the various types of plantation agriculture is bound up with the appearance of new economic wants, with new inventions, and with changes in fashion and diet. The tobacco plantation in Virginia took form with the acculturation of an Indian culture trait and the use of tobacco as a fashion on the European continent. Gillespie shows how the preference for sugar over honey in England during the sixteenth and seventeenth centuries led to the displacement of honey, the development of a large market for sugar, and the investment of English capital in West Indies sugar plantations.[27] The banana plantation became profitable with the coming of rapid communication, transportation, and refrigeration.[28]

The production of a certain specific commodity may enter into the establishment of a plantation as the original and determining purpose of those who invest the capital in the enterprise. Now that the developed pattern of the plantation is thoroughly a part of the capitalistic culture of America and Europe this is the case with most modern plantation settlements, such as the Ford and Firestone rubber plantations in Brazil and Liberia respectively. In earlier plantation colonies, however, the settlement frequently developed along economic lines not contemplated by those who sponsored it, and forced modifications in original intentions. This was true of Virginia. In such cases the particular staple, around whose production the plantation is organized, is a discovery subsequent to the establishment of the plantation itself. Concentration upon the culture of the particular agricultural commodity which proves most profitable leads to changes in its form and structure. It becomes a landed estate and an economic institution.

The peculiar features of this institution appear when agricultural industry is contrasted with manufacturing industry. Agriculture yields a different product when pursued on different kinds of soils and in different climates. Manufacturing industry, with the same skill and activity, may produce everywhere the same product. Agricultural production is dependent upon the seasons which means that it cannot be speeded up although its scale may be enlarged. The daily and minute division of labor which is characteristic of large-scale manufacturing industry is impossible in agriculture where the labor process is subject to constant interruption by seasonal changes. The seasonal nature of agriculture makes the requisite labor force imperative at critical periods. Where seasonal labor cannot be secured it becomes necessary to maintain sufficient labor throughout the year to meet such situations.[29] Again, the personal factor of management plays a greater role in agriculture than in manufacture since in agriculture the results of poor work may not appear until long afterwards. In manufacture the highly integrative nature of the division of labor is itself a check upon workmanship.

Large scale production in both agriculture and manufacture involves specialization of function within the industry but large scale agricultural production differs from the other in its greater dependence upon a uniform type of unskilled labor. This invariably leads to a sharp stratification between the common mass of workers and the relatively few who are above them.[30] The sharp distinctions between the two classes, together with the regulations which are enforced against the laborers, operates to lower the status of the latter. The competition now becomes economic as well as biological. In the heterogeneous racial situation which plantation settlement involves, competition effects a vocational distribution of population in which initial advantages and capacities for different kinds of work are the determining factors. Competition and selection result in a division of labor within the institution in which the proprietors, managers, and laborers represent, or tend to represent, different racial groups.

The Organization and Control of Labor

In the efficient management of his plantation in order to secure maximum returns, the disposition of the planter is to attempt to handle his laborers as so many separate machines, the value of each of which is determined by the quality and quantity of its product.[31] But the workers so regarded do not cease to behave as persons and to act in certain ways contrary to the best economic interests of the proprietor. The further development of the plantation is, therefore, in the direction of a type of control which will adjust its subordinated members to the particular tasks involved in the production of the staple crop.

In the state of racial and cultural diversity in which plantation settlement originates such control necessarily inheres in a system of relationships dependent upon the will or authority of a dominant individual or a class of individuals. In other words, the control is political and not social or cultural, that is, it does not inhere in common customs and traditions. So long as the frontier character of the settlement continues, the plantation is only slightly a part of a larger society whose *mores* would otherwise be compelling. So long as this is true there is a considerable degree of finality and independence in the authority of the planter over those resident upon his plantation comparable to the authority of a lord over his manor. Except for the necessary market relation the plantation is, to a high degree, isolated, and the central government is far away. The authority of the planter declines, or is taken from him, as the frontier passes upon the coming of better means of transportation and communication.[32]

Political control on the plantation is not merely for the purpose of maintaining order but for the purpose of coordinating labor in the work of agricultural production as well. It is necessary that the work be carried on efficiently and consistently. The internal integration of the plantation is not achieved, as in the case of the feudal manor, through the requirements of offense or defense in war, but in a competitive struggle wherein the plantation is set over against its market. Political control is achieved, therefore, not in conflict with the outside world, but in competition with it. For this reason the plantation as a political institution is subject to certain real limitations. The members do not see the enemy. They do not know those with whom they are in competition. The workers do not understand the market relation upon which they, in common with the planter, are dependent. Only the planter, in some greater or less degree, is able to envisage the whole situation and to have a well-defined end in view which entails, in addition to managing his labor, problems of production, of selling and of purchasing.[33]

If, therefore, political control undergoes change with the passing of frontier society it also changes with modifications in world market processes. The extension of communication and transportation enlarges the area of economic interaction and creates a new market situation to be adjusted to. This tends to break down, or at least puts a strain upon, old forms of labor control when the effort is made to carry them over into the new situation. Eventually it becomes necessary to adopt new forms of control. This may lead to the importation of new labor groups whose members can be treated more impersonally as utilities and perhaps held in slavery or to long terms of indenture. In the long run, however, forms of control tend to evolve which progressively take into account the workers' attitudes and wishes and allow them to participate in the control. But as this takes place the plantation is in process of breaking up into smaller units of proprietorship.

Peasant Proprietorship and Cultural Homogeneity

Oppenheimer has laid down the general proposition that "the ownership of large estates" is "the first creation and last stronghold of the political means," that is, the use, in some degree, of arbitrary control for economic gain. The large estate is "the last remnant of the right of war."[34] The goal of social and economic evolution is, however, toward what Oppenheimer terms a "freeman's commonwealth" based upon the small farm and democratic relations. Without assuming, as Oppenheimer seems to do, that such a community represents the final perfectibility of man and his institutions, it does seem to be true that peasant proprietorship is the last state in a cycle which begins with the large estate. This is not to say, however, that the cycle cannot begin over again.[35]

The transition from large estate to peasant proprietorship may take place by revolution, by legislation or by gradual change. In France, in Russia, in Roumania, in Hungary and to some extent in Mexico, it has resulted from revolutionary labor movements.[36] In such places as Denmark, Ireland and Virginia, the change has been more evolutionary.[37] But in either case it has tended to accompany a transition from extensive agriculture to relatively intensive agriculture. Peasant proprietorship marks the passing of the ruthless, exploitive methods of the frontier to concern for the conservation of natural resources when settlement has been completed and men have come to accept the new country as a homeland.[38]

How much the transition to peasant proprietorship is due to economic and ecological factors and how much to social factors is impossible to determine. They both interact to effect the final result. "The type of agriculture frequently depends not only upon transportation," says W. S. Culbertson in discussing the economic factors, "but upon the size of the agricultural unit. If agriculture is organized in a large way, and carried on in large farms, or estates, the nature of the crops is likely to be different from the case where small peasant farming is the rule. The breaking up of the large estates in Eastern Europe is . . . having a striking effect upon the character of agriculture."[39]

The rate of the change varies for different crops. Coffee, for example, does not require extensive production equipment. In addition, coffee farming is aided by the fact that even road and wagon transportation is not required; a burro is sufficient to carry the coffee bags to market. For these reasons cultivation by peasant proprietorship alongside, or following, plantation cultivation is in some areas a fairly rapid result. Cane sugar, on the other hand, requires heavy capital investment for the processing which must accompany production. Hence the sugar plantation tends to have a long life but peasant sugar farming is by no means impossible.[40] In any case where peasant commercial farming succeeds estate farming, economies in marketing tend to be secured through the cooperative pattern.

When peoples of different racial stocks are brought together in daily contact their relations are inevitably humanized. Masters of livestock as well as of men, in spite of temptations to act otherwise, grow fond of the creatures that are helpful and responsive to them.[41] Those who live and work together in a common situation come to regard themselves as "belonging" to each other and refer to themselves as "we." It is this that Park has in mind when he speaks of "forces inherent in the institution itself" which tend to modify it.

It has been said that slavery was doomed by forces inherent in the institution itself. This is probably true. It is probably true of every institution; but slavery in America was not permitted to survive until the forces which were destined to destroy or transform it had manifested themselves unmistakably. We are only now witnessing in the United States and in Brazil and the West Indies the slow working out of consequences that emancipation interrupted but could not profoundly alter. The slave was probably predestined to be what he has since very largely become, a peasant farmer.[42]

These "forces inherent in the institution" are not primarily economic but social forces, although they have economic effects. When the forces of personal and human relations begin to rise in and permeate the stratified community there is established a moral order, a common culture, and common values. In the interracial situation, which is the plantation situation, this result is facilitated by the intermediate physical type, the product of miscegenation, who tends to break down the caste system which arises in the effort of masters and employers to maintain the symbiotic and purely utilitarian relation between the races. A working accommodation or adjustment of dissimilar racial attitudes might persist indefinitely were it not for the presence of mixed-blood groups who, to some extent, participate in both worlds but do not completely belong to either. So situated they are able to interpret one group to the other, and the conflicts between the fundamental issues and interests of the two groups are subjectively registered in their own personalities as mental conflicts. For this reason the mixed bloods become the crucibles of cultural fusion and racial assimilation. A moral or sacred order resting upon the small farms thereupon grows up out of the secular as the final stage of the same process in which the plantation as an industrial and political organization grows up under the conditions of the frontier.

Notes

1. The Plantation as a Social Institution

1. Teggart, *The Processes of History*, 91.

2. Keller, *Colonization*; Roscher and Jannasch, *Kolonien, Kolonialpolitik, und Auswanderung*, 23–24; Leroy-Beaulieu, *De la Colonization chez les Peuples Modernes*, volume 2, 563–593.

3. Maine, *Ancient Law*, 129.

4. Teggart, op. cit., 79.

5. "These two theories have this in common, namely, that they both conceive civilization and society to be the result of evolutionary processes—processes by which man has acquired new inheritable traits—rather than processes by which new relations have been established between men" (Park, "[Human] Migration and the Marginal Man," 64).

6. Teggart, op. cit., 155.

7. Ibid., 49

8. Teggart, *Theory of History*, 196.

9. Teggart, *Processes of History*, 88–89.

10. The meanings of "open" resources and of "forced" labor are discussed later in the present chapter.

11. Roscher and Jannasch, op. cit., 23.

12. Data for the construction of a more accurate plantation map are not available. The first world agricultural census, taken in 1930, adopted no definition of the term "plantation" but grouped a considerable number of plants under the category "miscellaneous plantations" (*World Agricultural Census of 1930*, 49). In a letter to the writer dated October 25, 1930, Mr. H. Brizi, secretary general of the Institute, says: "I do not think that a map of the world showing plantation commodities exists, nor do I think it will be possible to construct a complete one with the data of this first world census, as such data are difficult to obtain for some countries and probably will only be obtained when the next census is taken."

13. Keller, op. cit., chapter 1.

14. Ibid., 8.

15. Ibid., 7.

16. Ibid., 10.

17. Ibid., 11.

18. "Colonial furnaces (in Pennsylvania) were described as baronial and patriarchal, resembling a feudal holding or a Southern plantation. They were located where forests were within easy reach, and generally had a farm adjacent; with the slaves, white servants, and free laborers, one of these furnaces formed a little settlement. Proprietors of furnaces clamored for Negro slaves and complained against the enlistment of redemption servants" (Herrick, *White Servitude in Pennsylvania*, 64).

19. Fuchs, "The Epochs of German Agrarian History and Agrarian Policy," 223–253.

20. The north and south trade is increasing in relative importance. It is the trade of the future, according to J. Russell Smith, *Industrial and Commercial Geography*, part 2, chapter 1.

21. von Engeln, *Inheriting the Earth*, 265–266.

22. Dodd, "The Plantation and Farm Systems in Southern Agriculture," 74.

23. Concerning Keller's distinction between farm colony and plantation colony, Isaiah Bowman states in a letter of October 18, 1926, to Robert E. Park: "The question of homestead vs. plantation . . . is undoubtedly important. . . . The limitation of the idea in a study of pioneer belts arises from the fact that the two types of land development may exist side by side. . . . The most important things . . . seem to me to lie back of the two contrasting systems that Keller recognizes rather than in the systems themselves, chiefly because these systems as they work themselves out in a given region vary so greatly from place to place . . . [that] . . . the regional differences in the plantation systems are as great as the contrast between that system and the homestead system. If we go into the homestead areas we shall find very much the same thing true. . . . So . . . I should say that the idea is a suggestive one but that its limitation are . . . more interesting and important than its applications."

24. Murray, *The Rise of the Greek Epic*, 55–57.

25. Turner, *The Frontier in American History*, 30, 37.

26. Macleod, *The American Indian Frontier*, 357–382.

27. Cited by Belaunde, "The Frontier in Hispanic-America," 213.

28. Moret and Davy, *From Tribe to Empire: Social Organization among Primitives and in the Ancient East*; Perry, *The Children of the Sun*, 129–131, 146, 445–447.

29. Nieboer, *Slavery as an Industrial System*.

30. Ibid., chapters 2 and 3.

31. Ibid., 174–177, 294–296.

32. Ibid., 420.

33. Ibid., 421.

34. Macleod, "Economic Aspects of Indigenous American Slavery," 640.

35. See Park and Burgess, *Introduction to the Science of Sociology*, 665.

36. Teggart, op. cit., 85.

37. Oppenheimer, *The State*, 27–81.

38. Ibid., 282.

39. The change to the new principle in Greece is illustrated in the poet Hesiod's advice to cultivate good terms with neighbors: "they are better friends than relatives, for if help is needed, a neighbor will come running in his shirtsleeves, while a relative will stop to put on his coat" (Quoted by Howard Becker, "Ionia and Athens," 33–34).

40. Sorokin, Zimmerman, and Galpin, *A Systematic Source Book in Rural Sociology,* volume 1, chapter 3, "Origins of Rural-Urban Differentiation."

41. Ripley, *The Races of Europe,* 29. See Park, *The Immigrant Press and Its Control,* 21–23, for evidence of this in the distribution of European languages.

42. Thomas and Znaniecki, *The Polish Peasant in Europe and America,* volume 4, part 2, chapter 6, "The Role of the Peasant in National Life."

43. Park and Burgess, op. cit., 798.

44. Oppenheimer, op. cit., 67.

45. "No matter in what quarter of the world we look, wherever there are native races one or more of these customs is practiced except where native customs have been destroyed by European influence" (Carr-Saunders, *Population,* 16).

46. "In any country at any given time there is a certain amount of skill and knowledge available and there are certain habits and customs which govern the use made of this skill and knowledge. Taking all these conditions into consideration, then it is clear that there is a particular density of population which must be reached and must not be exceeded if the largest possible income per head is to be obtained. . . . There is for any piece of land, when a certain amount of skill is available, a point when, by the application of a definite amount of capital and labour, the maximum return per head is reached; if less is applied the return per head will again be less, and if more is applied the return per head will again be less, though in this latter case the total produce will be greater. So in any country, however many complications may be introduced by the rise of industrialism and the exchange of manufactured articles for food grown abroad, there is a density of population which is more desirable than any other from the point of view of income per head. This may be called the 'optimum' density" (Ibid., 26–27).

47. "Capital thrives best in a settled order of society, where the risks of loss are at a minimum. It accepts favors from government, to be sure, but politics is no part of its game; peace, and freedom from disturbing innovations, are its great desiderata. Speculative enterprise, on the other hand, thrives best in the midst of disorder. Its favorite field of operation is the fringe of change, economic or political. It delights in the realm where laws ought to be but have not yet made their appearance. . . . Politics, thus, is an essential part of the game of speculative enterprise" (Johnson, "The War—by an Economist," 420–421).

48. "It somehow shocks the sense of fairness of hard-headed white and yellow people that semi-savages should be driving ill-bred sheep, scraggy cattle or ponies hardly fit for polo over plains and mountains that are little less than great treasure-vaults of valuable minerals and chemicals; or that they should roam with their blow-pipes and bows and arrows through forests of inestimable value for their timber, drugs, dyes, lattices, gums, oil-seeds, nuts or fruits; be turning this wealth to no use, nor allowing it to circulate in the world's markets" (Johnston, *The Backward Peoples and Our Relations with Them,* 59).

49. See Macleod, "Big Business and the North American Indian," 480–491.

50. The following was written by an adventurer who attempted to establish a coconut plantation on a small South Sea Island. After experiencing a good deal of difficulty in clearing the land and in getting a native tribe to work, his laborers suddenly decided to stop work altogether. It was just when victory was approaching that they announced their intention of going away.

"Must go for walk-about," one explained shortly when, aghast, I offered remonstrance.

"Must have spell from hard work too much," grunted another.

I stormed at them, and threatened, saying that if they went I would never employ them again, that I would bring labourers from some other part of the coast. They made no answer, but stood staring sullenly, one or two muttering, and some so fingering their spears that I though it best to let them see I had my revolver in my pocket, as I always had.

Then I spoke to them softly, and, pointing to the plantation, begged that they should not leave me just as the fruit of our recent hard laborings was in sight; and asked had I not been a good master to them, giving them in plenty of the good things in my store; and drew their attention to the fact that if they went away they would have none of these good things, and would miss them very much.

And still they stood there, a sullen, silent crowd, the women by gestures hushing their children, the men scowling up at me, the dogs sniffing in and about the array of thin, black legs.

Then I turned to Mary Brown, who had come from the kitchen, and stood behind me at the head of the veranda steps. I told her to point out to them in their own language and in her own way the folly of their leaving me, to use all the arguments she knew to induce them to stay, to do her best for me, as often she had done her best for me before. Then I stood back and made room for her to speak to the sullen, silent crowd on the ground.

But Mary, who had never failed me before, failed me now. She made no speech to those on the ground, but looked at me earnestly a while, and hesitatingly and stumblingly said that they would have to go—and that she should have to go with them.

It was the fashion for natives to go for 'walk-about,' she said, in the manner of one who excuses, but does not intend to allow the excuse to interfere with a fixed determination. Always had they been like that. Always, she repeated. With white men it was different. White men could stay long times in one place. But with natives, their eyes came tired of looking always at the same things. And their feet came tired walking always on the same ground. And their bodies came tired from sleeping always in the same camp. They wanted to look at other places, walk on other ground. They would have to go. Yes. They would go to a place three days' walk down the coast. And she would go with them, because she was one of them, and felt the same as they did—though her heart was sore that she should leave me, and the heart of Willie, her husband, was sore too, and she would like to stay, only . . . only . . .

I watched them out of sight, and, till the sinking of the sun was done, stared at the empty beach. Then I looked at the deserted camp and the wavering smoke of one of its dying fires, and at a rain squall sweeping up with a great rustling of leaves, and around my quiet and lonely house; and I won-dered if, after all, there was anything so tremendously self-satisfying in being a Settled and Respectable Person" (McLaren, *My Crowded Solitude,* 94–97).

51. Merivale, *Lectures on Colonization and Colonies,* lecture 11, 304 ff. The habit of steady industry which the Barbadian Negro has been forced to maintain has given him a reputation for thrift throughout the Caribbean area, whereas the Haitian and Jamaican Negroes, in those two islands, are reputed lazy and worthless for plantation labor. "A rich field of investigation awaits the sociologist," says J. Russell Smith, "who will hasten to study the Haitian negro before American control controls him too much and compare him with his Barbadian brother, who considers himself the aristocrat of the West Indies. He boasts of the fact that Barbados has never been anything but a British possession. He is proud of his neat home and his perfectly cultivated bit of earth. Two-thirds of his chil-dren between the ages of five and fifteen are in school" (*North America,* 739). Transplanted to other plantation areas, however, the Haitian and Jamaican Negroes seem to work well enough under stringent control. The investment of American capital in plantations in the West Indies is distributing them over the Caribbean area, where their transportation is referred to as the slave trade, (Beals, "The Black Belt of the Caribbean," 129–138).

52. Park, "The Urban Community as a Spatial Pattern and a Moral Order," 4.

53. Cf. Teggart, *Theory of History,* chapter 12, "Events in Relation to the Study of Evo-lution."

54. Clements, *Plant Succession and Indicators.*

55. In the West Indies may be found the end of one cycle and the beginning of another. The small farms in some of the British islands, formerly the scene of large plantations, are assisted by cooperative marketing, rural credits, agricultural extension education and other measures fostered by the government. In the islands owned or dominated by the United States the highly efficient plantation with modern machinery and methods operates with imported Negro labor from Haiti and Jamaica, (Beals, op. cit., 132).

56. "Civilizations are the most external and artificial states of which a species of devel-oped humanity is capable" (Spengler, *The Decline of the West,* 31).

57. Park, "[Human] Migration and the Marginal Man."

58. In Murray's *New English Dictionary* a number of uses of the term plantation are given. Perhaps these are due to the wide meaning assumed by the infinitive *to plant* and its extensive use by the people of medieval England. The most obvious use of the word denoted the placing of plants in the soil that they might grow. It was in this sense that Aaron's rod was described by a writer in 1450 as having "fructified without plantacioune." It was easy to extend this use to the act of establishing or founding anything as, for exam-ple, a religion. Thus Bacon speaks of "those instruments which it pleased God to use for the plantation of the faith." Or it might refer to a settlement of people as in Coverdale's Bible translation: "I will appoynte a place and will plante (Vulgate *plantabo*) them, that they maye remayne there." With this use of the verb the noun might refer to a group migrating or being transported (and transplanted): "Ascanium . . . carrying forth a plan-tation of men . . . found a white sow with 30 pigges sucking her" (952–953).

59. Francis Bacon, "On Plantations," volume 2, 194.

2. The Metropolis and the Plantation

1. McKenzie, "Food Supply in Relation to Population," 175.

2. For the coming and meaning of this geographical revolution in England, see G. B. Parks, *Richard Hakluyt and the English Voyages,* chapter 1, "The New Geography in Tudor England."

3. Shepherd, "The Expansion of Europe," 43.

4. "Instead of resting chiefly on the basis of an exchange of products of the soil and on an exchange of those brought forth by a narrowly local and restricted handicraft, the economic system of the continent has come to be founded upon actual money in silver and gold. It has come to rest also, on an exchange of the most varied products fashioned by the skill of mankind everywhere on earth; and this exchange has been reckoned in terms of money. The financial revolution, moreover, has brought on tremendous fluctuation in the value of money, and hence in prices, along with an inevitable disarrangement of the pecuniary standards of living" (Ibid., 222). See also Knight, Barnes, Flugel, *Economic History of Europe,* 307–314.

5. Op. cit., 348–362. Nieboer contends that the liberation of the serf was primarily due to the appropriation of all land, the increase of population, and the closing of resources. The commutation of services paralleled, but was not identical with, the transition from serfdom to freedom.

6. Hammond and Hammond, *The Rise of Modern Industry,* 81–96.

7. Beer, *The Origins of the British Colonial System, 1573–1660,* 44–45, 51.

8. "This community of Atlantic nations has as yet no name. Its essential unity has not yet been recognized sufficiently clearly for that and it has been torn by family quarrels. It is not an empire, for that suggests unity of rule as well as of culture. We may perhaps call it a 'commonwealth,' the Commonwealth of the Atlantic—for the weal of each nation in it affects all the others" (Traquair, "The Commonwealth of the Atlantic," 606).

9. "A study of oceanic commerce up to the nineteenth century would show, in all probability, that, although relatively considerable, it was much smaller in amount than has been supposed. Whatever the influence of the course of expansion on Europe in other respects, along commercial lines, at least, it seems to have been of comparatively slight significance for several hundred years. It was not much before the nineteenth century that the ultimate effects of overseas trade became very distinctly perceptible." "That there was a commercial revolution is unquestionable; but up to the nineteenth century it had to do far more with the potentialities arising out of change in the highways of commerce than with commerce itself. In this sense a 'revolution' may be said to have occurred" (Shepherd, op. cit., 60, 218).

10. See Hammond, op. cit., 97–189.

11. The importance of cheap transportation in England's industrial revolution was stressed by Alfred Marshall in the following significant statement: "Probably more than three-fourths of the whole benefit [England] has derived from the progress of manufacturing during the nineteenth century has been through its indirect influences on lowering the transport of men and goods, of water and light, of electricity and news; for the dominant economic fact of our own age is the development, not of the manufacturing, but of the transport industries. It is these that are growing most rapidly in aggregate volume and

individual power. . . . They also [are the industries] which have done the most toward increasing England's wealth" (*Principles of Economics,* 674–675).

12. "Beginning in the North and going round the compass the companies were as follows: the Eastland Company, trading to Scandinavia and the Baltic; the Russia Company; the Merchant Adventurers, controlling the trade from Denmark to France, where the free-trade gap appears; the Levant Company, trading in the Mediterranean; the Guinea or Africa Company; the East India Company, with its immense Asiatic field; and then the various companies familiar to the students of American History, the Virginia Company, the Plymouth Company, later the Hudson's Bay Company, etc. By means of the trade of these companies England marketed her surplus wares, especially her woolen fabrics, and imported the goods of which she stood in need—naval stores from the Baltic, manufacturers and wine from the Continent, gold from Africa, oriental products, and furs and fish from America" (Day, *A History of Commerce,* 204).

13. *Some Account of the Province of Pennsylvania* (1681).

14. Rose, Newton, and Benians, *The Cambridge History of the British Empire,* volume 1, 93.

15. *Romance,* 35.

16. Mackinder, *Democratic Ideals and Reality.* Realities are the forces and factors "that have conditioned History, and have led to the present distribution of population and civilization" (Ibid., 111, footnote).

17. Ibid., 81. Actual discussion of this land mass, however, treats it as a World-Promontory stretching southward from a vague, inaccessible, and uninhabited north.

18. Brunhes and Vallaux, *Le Géographie de L'Histoire,* 120–192. Quoted in Dawson and Gettys, *An Introduction to Sociology,* 150–155.

19. The trading factory, or post, is a business establishment in charge of factors or agents in a foreign country. The Italian factories, or *fondachi,* in the cities of the Levant during the period of the Crusades became models for later European factories in foreign countries (See Keller, op. cit., 63 ff).

20. Op. cit., 141.

21. Ibid., 145–146.

22. Supported by the Crusaders, the Venetians acquired lands outside the cities of the Levant in which they had factories and converted these lands into large estates under the charge of overseers (Keller, op. cit., 64).

23. Lucas, *The Beginnings of English Overseas Enterprise: A Prelude to the Empire,* 60.

24. Former companies, with the exception of the Muscovy Company, were "regulated" companies in which the members enjoyed the safeguards of organization but individually traded on their own account.

25. Knowles, *The Economic Development of the British Overseas Empire,* 9.

26. Dryer, "Mackinder's 'World Island' and Its American 'Satellite,'" 205–207.

27. Mooney estimates on the basis of indications during some early stage of European contact, that there were only 846,000 Indians within the limits of what is now the United States about the time of first settlement by Europeans. Of these the densest Indian settlements were along the Pacific coast. In about 1600 the present states of New England, New York, New Jersey, and Pennsylvania had an estimated Indian population of 55,600. The area which is now Maryland, Delaware, Virginia and the Carolinas had only 52,200 in

1600 (Mooney, *The Aboriginal Population of America North of Mexico*). Spinder thinks these estimates too conservative (Spinder, "The Population of Ancient America," 641–660).

28. Edited by E. Goldsmid, Edinburgh, 1885–1890.

29. *Past and Present*, book 4, chapter 3.

30. Hakluyt, op. cit., volume 8, 98.

31. Ibid., 111.

32. Ibid., 140.

33. Ibid., 117.

34. Ibid., 348–386.

35. Ibid., 441–442.

36. Ibid., 138.

37. Davis, *Corporations*, volume 2, 157.

38. We have here a type of settlement whose relation with the metropolis was not culturally *derivative* merely, as with the Greek colony, nor yet simply politically *related* to the mother city, like the Roman military colony, but one economically *connected* in a vital interdependence with a center which was primarily a market. Englishmen promoted this form of migration because it was expected to extend the community division of labor by producing new goods for the market. Such expectations were not always realized, of course.

39. Scisco, "The Plantation Type of Colony," *The American Historical Review*, 267–268.

40. "Out of the ruins of plantation efforts arose a modified form of remarkable vitality, that is to say, the New England town. In its completest form it was a corporate plantation, with combined powers of jurisdiction and proprietorship, and a small measure of economic unity" (Ibid., 269).

41. Sumner, "Advancing Social and Political Organization in the United States," 314–315.

42. Secretary Nichols stated that the exports of New England "maintain and supply the plantation of Barbados and Jamaica" (Cited in Wertenbaker, *The First Americans, 1607–1690*, 54).

43. In early Kentucky "the female sex, though certainly an object of much more feeling and regard than among the Indians, was doomed to endure much hardship and to occupy an inferior rank in society to her male partner; in fine our frontier people were much allied to their contemporaries of the forest in many things more than in their complexions" (Butler, *A History of Kentucky*, 135).

44. Mead, "The Philosophies of Royce, James and Dewey in Their American Setting," 214.

3. The Plantation in Virginia

1. Bruce, op. cit., volume 1, 74–75.

2. Smith, *Works*, 148.

3. Beverly, *The History of Virginia*, 138.

4. Gee and Corson, *Rural Depopulation in Certain Tidewater and Piedmont Areas of Virginia*, 5–6.

5. Macleod, *The American Indian Frontier*, 176.

6. Macleod, "Debtor and Chattel Slavery in Aboriginal North America," 375. In the Carolinas, it is interesting to note, the term for slave meant, "that which is obsequiously to depend on the master for food," and was applied alike to pets, domestic animals, and slaves (Lawson, *History of Carolina*, 327). It is evident that slavery among the Southeast Indians was not a thoroughly established institution.

7. Beverly, op. cit., 195.

8. Kingsbury, "A Comparison of the Virginia Company with the Other English Trading Companies of the Sixteenth and Seventeenth Centuries," 168.

9. Ibid., 166–167.

10. Jefferson, *Notes on the State of Virginia*, 153.

11. Bruce, op. cit., volume 1, 491.

12. Ibid., 498–499; Macleod, *The American Indian Frontier*, chapter 4.

13. Doyle, *English Colonies in America*, 128.

14. Bassett, *A Short History of the United States*, 46

15. John Fiske, for example, says: "With statesmanlike insight he [Dale] struck at one of the deepest roots of the evils which had afflicted the colony. Nothing had done so much to discourage steady labour and to foster idleness and mischief as the communism which had prevailed from the beginning. The compulsory system of throwing all the earnings into a common stock had just suited the lazy ones. Your true communist is the man who likes to live on the fruits of other people's labour. If you look for him in these days you are pretty sure to find him in a lager beer saloon, talking over schemes for rebuilding the universe. In the early days of Virginia the creature's nature was the same, and about one-fifth of the population was thus called upon to support the whole. Under such circumstances it is wonderful that the colony survived until Dale could come and put an end to the system" (*Old Virginia and Her Neighbours*, volume 1, 166).

16. Macleod. op. cit., 177.

17. Macleod, "Big Business and the North American Indian," 460–491.

18. "He found on arrival that no crops were planted, although the planting season was past. The men's chief occupation was bowling in the streets, the houses were falling in pieces, and the Indians were defiant. He turned on Newport, who had continually deceived England about the state of the colony, pulled his beard in public, threatened to hang him, and asked 'wheather it ware meant that the people heere in Virginia shoulde feede upon trees.' He set the colonists to digging sassafras roots and hewing cedar for the profit of the Company. The spiritless inhabitants did not resist, but fled to the woods: when he took them he burned them at the stake. For stealing food some were hanged, and one was tied to a tree to starve" (Bassett, op. cit., 49).

19. Bruce, op. cit., volume 1, 212.

20. Craven, *Soil Exhaustion as a Factor in the Agricultural History of Virginia and Maryland, 1600–1860*, 59–60. It is evident that land was virtually free or at least very cheap. Under the strict application of the principle of headright it cost no more than a shilling an acre since the transportation charge on an immigrant was £6. In 1705 the practice of direct sale of Crown lands was sanctioned; fifty acres might be had for five shillings under certain easy conditions. Until the Revolution, of course, the Crown was recognized

as the ultimate proprietor of all lands in Virginia, and the immediate owners paid a quit-rent of a farthing an acre. The planters protested paying this feudal tribute to the king, and the squatters on the frontier occupied land which they came to regard as their own and for which they refused to pay a quit-rent (Coman, *The Industrial History of the United States,* 32, ff).

21. Bruce, op. cit., volume 2, 524.

22. Ballagh, *White Servitude in the Colony of Virginia,* 22–23.

23. Lauber, *Indian Slavery in Colonial Times within the Present Limits of the United States,* 288–289. Inability to control the intractable adult Indian in his native haunts led the planters to prefer Indian children because of the greater probability that they might be trained to usefulness. "The grown persons of the race . . . were in many cases unmanageable, and hardly worth the constant attention required to control them" (Bruce, op. cit., volume 2, 54).

24. "So important are the effects of soil and climate on the quality of the tobacco produced that even in those countries which, as a whole, grow a product of relatively low grade, tobacco culture is more or less definitely localized" (Buechel, *The Commerce of Agriculture: A Survey of Agricultural Resources,* 343).

25. Ibid., 340.

26. "It was not very long before a certain place on the James River acquired the name Varina from the supposed similarity of the tobacco produced there to the celebrated Spanish Varinas (Bruce, op. cit., volume 1, 218).

27. *Cal. of State Papers: Colonial, 1631.* Quoted in MacInnes, *The Early English Tobacco Trade,* 137.

28. Wertenbaker, *The Planters of Colonial Virginia,* 25.

29. Ibid., 64.

30. Wertenbaker, op. cit., 69.

31. Schmidt and Ross, *Readings in the Economic History of American Agriculture,* 89.

32. The figures in Table I do not represent the entire export trade. Customs duties were becoming so heavy that smuggling on an extensive scale was resorted to.

33. "A factor may be defined as an agent empowered by an individual or individuals to transact business on his or their account. Usually he was not resident in the same place as his principal, but in a foreign country or at a distance. The business which he transacted depended on the authority given him by his principal, and might be limited to a particular and specific transaction, or might be more extended, and comprise buying and selling, shipping, negotiating insurance, discounting bills, making payments, etc. . . . A distinctive mark of the factor seems to have been that he was permitted to transact business in his own name and as if on his own account" (Buck, *Development and the Organization of Anglo-American Trade, 1800–1850,* 6–7).

34. Macpherson, *Annals of Commerce,* volume 3, 163.

35. McKenzie, "Spatial Distance," 536–544.

36. The most important study of agriculture from this point of view is by J. H. von Thünen, *Der isolierte Staat in Beziehung auf Landwirtschaft und Nationalökonomie.* See Othmar Spann, *The History of Economics,* 171–187. Thünen assumes a large market town in the middle of a fertile plain with homogeneous soil, climate, and transport conditions;

only the factor of distance from the market is left variable. The greater the distance from the market, the greater the cost of transportation. Thus the various types of agriculture and horticulture will be arranged in concentric zones around the market city as follows: Zone I, Horticulture and market-gardening; Zone II, Sylviculture; Zone III, Cereals; Zone IV, Stock-raising; Zone V, Hunting. From the theoretical assumptions under which Thünen worked, his conclusions are, of course, not expected to fit all empirical situations but they do describe agriculture under the limited transport conditions of the Middle Ages. With modern refrigeration and faster and cheaper means of transportation, distance relations are greatly changed; South America, for example, can supply Germany with grain. But remembering that Thünen's distance from the market is based upon cost and not upon geographical space, his concentric circles continue, in large measure, to hold true. Thünen's studies show that the advantage of one type of agriculture over another in a region is relative, not absolute.

37. Buechel, op. cit., 343. "As to other commodities producible here, as pipe-staves, timber works of all kinds, and corn . . . our position here is so remote that the cost of freights and transport devours the whole produce" (Colonel Nicholas Spencer to Sir Leoline Jenkins, May 13, 1681, *Calendar of State Papers, Colonial Series, American and West Indies, 1691–1685,* 47).

> Tobacco had a great advantage over all the other agricultural products of Virginia in the fact that it could be produced in larger quantities to the acre. This was of supreme importance in a country where so much labor and patience were required to clear the ground of its primaeval growth in preparation for planting or sowing. Tobacco, moreover, could be shipped to England in more valuable bulk to the space than any other agricultural product. As a result of this circumstance, the pecuniary return upon a cargo of it was larger than upon a cargo of any ether commodity of the same general nature in proportion to the expense of transportation for so great a distance. Tradition and habit doubtless brought to bear a strong influence in the subsequent history of Virginia to promote the cultivation of tobacco, but in the beginning it was an economic necessity, and in no small degree it continued to be such (Bruce, op. cit., volume 1, 260–261).

38. E. W. Burgess, "The Growth of the City."

4. Plantation Management and Imported Labor in Virginia

1. Wertenbaker, op. cit., 29. With the discovery that Virginia could produce commercial tobacco a relatively heavy migration set in, following 1618.

2. Ibid., 32

3. Ibid., 34, ff.

4. Marcus W. Jernegan, "A Forgotten Slavery of Colonial Days," 746.

5. Paul H. Douglas, "American Apprenticeship and Industrial Education," 234–235.

6. Quoted in Bruce, op. cit., volume 2, 1–2 footnote.

7. Ballagh, op. cit., 33.

8. Jernegan, op. cit., 749.

9. Ballagh, op. cit., 42–43.

10. Ballagh. op. cit., 43–44.

11. Ibid., 45.

12. Quoted in ibid., 50–51.

13. Ibid., 51.

14. Ibid., 58.

15. Ibid., 45.

16. Ibid., 58–59.

17. Quoted in ibid., 60

18. An opposite trend was taking place at the same time in the colony north of Virginia and Maryland, Pennsylvania. Pennsylvania was predominantly a colony of manufacturing industry, as Virginia was of agricultural industry. The demand for skilled labor in Pennsylvania tended to eliminate the Negro slave and increased the importation of indentured servants. This undoubtedly accounts, in large measure, for Pennsylvania's opposition to Negro slavery. Herrick says: "In all the anti-slavery discussion in America, Mason and Dixon's Line was the Southern boundary of Pennsylvania. That the line of division between slavery and freedom was drawn south of Pennsylvania was due in no small measure to the substitute form of servile labor which was available to this state" (Herrick. op. cit., 285).

19. *The Cambridge History of American Literature* (Cambridge, England, 1917), volume 1, 14.

20. This is one reason why primogeniture and entail failed to gain a foothold in the transmission of plantation large estates through inheritance, although these English practices were instituted in all the American colonies. "In Virginia the old lands were continually wearing out and new lands had to be secured. That meant alienation, or at least abandonment, of one tract and the purchase or patent of another. It meant movement from the older to the newer sections, one of the characteristic features of much of the South before the Civil War. Entail was designed for a fixed estate, not a varying one. It was unsuited to the needs of a class who continually found their old lands wearing out and at the same time could secure fresh lands for next to nothing. Such a situation was far different from that of the English cousin who had a fixed estate settled upon which, perhaps, had neither decreased nor increased by an acre for centuries. As in many other respects, Virginia was a replica of England in legal form of land tenure, but in fact widely different, due to circumstances such as given herewith" (Keim, "Influence of Primogeniture and Entail in the Development of Virginia," 154).

21. The American slave-labor regime was developed, under a money economy, to enable European settlers and capitalists to exploit American resources with the aid of African labor. Any fixing of laborers to the soil as in serfdom, would have hindered the purpose in America. Success in the early task of conquering the wilderness and developing the varied opportunities required that those who controlled labor should be able to carry it from district to district, change it from occupation to occupation and transfer it at will to fresh employers. Accordingly, in the colonial regime which was inherited and further elaborated by the antebellum South, the laboring force was organized for quick response to either the

regional or occupational call of industrial opportunity. By reason of the slave-labor system the expansion of settlement in the South was actually far more rapid than in the North; and the development of new industries for which the regime was suited, cotton and sugar production for example, was accomplished with great speed. This mobility of labor was secured in part by the migration of planters and their shifting the employment of their slaves from staple to staple or from agriculture to handicrafts as the case might be. In very considerable part also it was attained through the services of the slave trade" (Phillips, "The Economics of the Slave Trade: Foreign and Domestic," 84–86). See also Henry Bolingbroke, *A Voyage to the Demerary,* 84–86; quoted in Phillips, *Plantation and Frontier Documents,* volume 1, 49–51.

22. "In interpreting into mental terms the consequences of gregariousness, we may conveniently begin with the simplest. The conscious individual will feel an unanalysable, primary sense of comfort in the actual presence of his fellows, and a similar sense of discomfort in their absence. It will be obvious truth to him that it is not good for the man to be alone. Loneliness will be a real terror, unsurmountable by reason" (Trotter, "Herd Instinct," 24).

23. See the case in Bruce, op. cit., volume 2, 110.

24. For example, Beth, the old servant in an English family, in E. F. Benson, *Mother* (New York, 1925).

25. It is interesting to note in this connection that in Roman law freedmen might, on complaint of their patron, be re-enslaved on the ground of ingratitude (Buckland, *The Roman Law of Slavery,* 422–424).

26. On the basis of early contacts between the races in other New World settlements it is probable also that there were olfactory and hygienic causes of cleavage. Du Pratz in Louisiana advised concerning precautions to take with Negroes as follows: ". . . that you may be as little incommoded as possible with their natural smell, you must have the precaution to place the negro camp to the north or northeast of your house, as the winds that blow from these quarters are not so warm as the others, and it is only when the negroes are warm that they send forth a disagreeable smell. The negroes that have the worst smell are those that are the least black; and what I have said of their bad smell ought to warn you to keep always on the windward side of them when you visit them at their work." Le Page Du Pratz, one of the first Louisiana planters, wrote his book on "The Negroes of Louisiana; of the Choice of Negroes; of their Distempers, and the Manner of Curing them" to give other planters in the region the benefits of his experience. Quoted in Saxon, *Old Louisiana,* 72.

27. Russell, *The Free Negro in Virginia, 1619–1865,* 11.

28. Ballagh, *A History of Slavery in Virginia,* 29. It thus appears to have been a practice to separate Negroes even before they legally became slaves. The separation of Negroes belonging to the same tribe and who could understand each other was a common practice during the heyday of the foreign slave trade. It was both an incident of purchase by different buyers and a deliberate policy designed to prevent insurrection. The result was to tear the Negro thoroughly apart from old group relations, interrupt the transmission of tribal traditions, and present him disorganized enough to rapidly take on the culture of the group into which he was introduced in so far as he was able and allowed to participate in

it. How rapidly Negroes were able to assimilate white culture in spite of their handicaps and how rapidly they lost their own is indicated in the following statement by the author of a recent study of a South Carolina Island group of Negroes: "Although merchants, in selling newly imported Negroes invariably advertised the tribe or geographical section from which the Negroes came, it is interesting that none of these facts were mentioned in selling seasoned slaves. It would seem that tribal differences tended to disappear as the slaves became seasoned" (Guion Griffis Johnson in T. H. Woofter, Jr., *Black Yeomanry*, 22). See Park, "Education in Its Relation to the Conflict and Fusion of Cultures," 38–63, for a discussion of the evidence for the almost complete break in the Negro's African tradition.

29. Russell, op. cit., 24.

30. Ibid., 32–33.

31. Other cases of Negroes who served out their time and became owners of land and masters of servants are given by Russell. Richard Johnson, for example, came in 1651 either as a free man or as a servant indentured for only three years, for three years after his arrival he received one hundred acres of land for importing two other persons (Ibid., 25–26). On the other hand, there are instances of Negroes serving unusually long terms, or at least recorded as indentured for long terms. John Q. Hamander was indentured for ten years. Two daughters of Emanuel Dregis were indentured, one for thirteen years and the other for twenty-nine years, but their releases were afterwards purchased by their father (Ibid., 26–27).

32. Ibid., 29

33. Ibid., 29–30.

34. Op. cit., book 4, 35.

35. Russell, op. cit., 18.

36. Ballagh, op. cit., 38–39.

37. Ibid., 44–45. The text of the act seems to uphold this view. "Whereas some doubts have arisen whether a child got by an Englishman upon a negro woman should be free or slave, be it therefore enacted by this present grand assembly, that all children born in this country shall be bond or free according to the condition of the mother, and if any Christian shall commit fornication with a negro man or woman, he or she so offending shall pay a fine double the fine imposed by the previous act" (Henning, *Statutes at Large of Virginia*, volume 2, 280).

38. Ibid., volume 1, 146.

39. In Maury, *Memoirs of a Huguenot Family*, 349–350.

40. Because of the strong white democracy which has developed in the Southern states especially since the Civil War it is somewhat difficult to comprehend the full meaning of social differences in which race questions figured relatively little in the early colonial South. The stratification was one of class rather than race. In Virginia, Humphrey Chamberlaine, of good birth, was fined even though he apologized because he had not shown due respect to the person of Colonel Byrd (Bruce, *Social Life of Virginia in the Seventeenth Century*, 133).

41. Wertenbaker, op. cit., 122–123.

42. "There were certain particular designations to show calling which were applied generally without social discrimination. In one instance alone, perhaps, did such a designation

carry a distinct inference of social importance without, however, nicely defining its degree; the word 'planter' probably at a late period conveyed such a meaning. Not long after the abolition of the Company, we find the term 'planter' applied to the lessees whose names appear in the grants of land belonging to the office of governor. The area contained in these grants was not extensive, and the lessees were men of no social consequence. Not many years later the term 'planter' was applied with great freedom, whether the patentee acquired title to a large tract or to a small; whether he was a citizen of marked prominence in the colony, or possessed no prominence at all. But by 1675, we find in the ordinary conveyances, recorded in the county courts, an indifferent use by the same man, as applicable to himself, of the terms 'gentleman' and 'planter,' as if the two were practically interchangeable. At this time, the estates, in many cases, spread over many thousand acres, and whilst all who owned and cultivated land of their own, whether great or small in area, were, in strict sense, planters, the term may have come to have a subordinate social meaning as applicable to men of large estates, whose social position by forces of birth, as well as of worldly possessions, was among the foremost in the community. Or it may be, which seems, on the whole, more probable, the person drawing up one of these deeds designated himself there as 'gentleman' if he happened at the moment to think of his social rank, or as a 'planter' if he thought of his calling." "Sometimes, in one deed, a grantor will designate himself as 'gentleman'; in a second as 'planter'; and in a third as 'merchant' . . ." (Bruce, op. cit., 110–112).

43. "Of the five hundred persons alive in Virginia in October 1609, all but about sixty had died by May of the following year" (Fiske, op. cit., volume 1, 154). Quoted in Wertenbaker, *Patrician and Plebeian in Virginia, or The Origin and Development of the Social Classes of the Old Dominion*, 8.

44. Ibid., 28–29.

45. Tschan, "The Virginia Planter, 1700–1775," 8.

46. *Virginia Magazine of History*, volume 3, 167.

47. Wertenbaker, op. cit., 17–18. Samuel Mathews "had married the daughter of Sir Thomas Hinton, the son-in-law of Sir Sebastian Harvey, one of the most distinguished Lord Mayors of London in those times" (Bruce, op. cit., 52). "Adam Thoroughgood . . . was a brother of Sir John Thoroughgood . . . who was attached to the Court" (Ibid., 52). It is evident that, commoner or nobility, those who were able to establish themselves as large planters were those able to command capital from some source. The controversy over the origin of Virginia's First Families would seem to be reduced to this.

48. Quoted in Wertenbaker, op. cit., 31.

49. Ibid., 58–59. With this change in the characteristics of the planter in the seventeenth and eighteenth centuries, Wertenbaker traces corresponding changes in attitude with respect to dueling, military achievements, and chivalry. These were not highly regarded with the merchant-planters but developed later.

50. Tschan, op. cit., 20.

51. Ibid., 5.

52. See *supra*, 65–66. "The word 'colony' merely took on a broader meaning than before, while 'plantation' remained what it had been, a local community subject to colonial government" (Scisco, op. cit., 268).

53. Ingle, "Local Institutions of Virginia," 28.

54. Ballagh, op. cit., 81. Later acts of 1788 and 1850 practically removed entirely, at least legally, the master's power of life and death (Ibid., 81–82).

55. Westermarck, *The Origin and Development of the Moral Ideas,* volume 1, 708.

56. "The [Roman] empire was an assumption by the state of functions and powers which had been family powers and functions, and part of the patria potestas. Women, children, and slaves shared in emancipation until the state made laws to execute its jurisdiction over them. Hadrian took from masters the power of life and death over slaves" (Sumner, *Folkways,* 289).

57. For a good illustration of this, see the account of a slave auction in Frederick Bancroft, *Slave-Trading in the Old South,* 109–110.

58. Tschan, op. cit., 21–23.

59. Bruce, *Economic History of Virginia in the Seventeenth Century,* volume 2, 123.

60. Munford, *Virginia's Attitude toward Slavery and Secession,* 118. Chapters 16 and 17 of this reference contain a number of representative specimens of deeds and wills of Virginia slaveholders emancipating slaves between 1782–1860.

61. Table III.

62. *Richmond Enquirer,* October 8, 1809, quoted in Russell, op. cit., 173–174.

63. Phillips, *Life and Labor in the Old South,* 243.

5. The Plantation and the Frontier

1. Beverly, op. cit., 283.

2. Craven, op. cit., 161.

3. Compiled from sources by C. M. Destler, "The Tobacco Industry in Virginia, 1783–1860," 9, 42–43.

4. Beverly, op. cit., 256.

5. Craven, op. cit., 66.

6. Gee and Corson, op. cit., 7.

7. Cited by Destler, op. cit., 25.

8. Craven, op. cit., passim.

9. Quoted in Wertenbaker, op. cit., 145.

10. Eighty thousand other slaves during the same period, according to the estimate, emigrated with their masters. *Slavery and the Internal Slave Trade in the United States,* 12, 15, 17; The *Virginia Times,* quoted in the *Niles' Weekly Register,* volume 51, 83. Professor Dew of William and Mary College, a contemporary writer, said, "The only form in which it can be said that slaves on a plantation are profitable in Virginia, is the multiplication of their number by births. . . . The value of the slaves so added to this number is the certain rice for which they will at any time sell to the southern trader. That master alone finds productive value in his increase of slaves, who chooses to turn the increase of this capital, at regular intervals, into money at the highest market price" (*Review of the Debate in the Virginia Legislature in 1831–1832*).

11. Ford, *The Writings of George Washington,* volume 2, 256.

12. Conway, *George Washington and Mount Vernon,* volume 4, 90.

13. Craven, op. cit., 121.

14. Ibid., 142.

15. Fleming, "The Slave-Labor System in the Ante-Bellum South," 117.

16. Craven, op. cit., 160–161. The following table compares the average size farm in Virginia with the average size farm for the whole United States since 1850 until 1925. It will be noticed that since 1900 Virginia's average size farm has been smaller than the average for the United States. A complicating factor, however, arises out of the fact that the Census Bureau includes in its definition of a farm, land operated by tenants, renters, or managers as well as owners. Thus a plantation with five tenants would count as five farms even though it is under unified ownership and supervision. There is no doubt, however, that the trend toward the small farm begun before the Civil War has continued since.

Census	Virginia	U.S.
YEAR	ACRES	ACRES
1850	340	203
1860	324	199
1870	246	153
1880	167	134
1890	150	137
1900	119	146
1910	106	138
1925	89	145

From the *University of Virginia News Letter* 6 (February 1, 1930).

17. Knowles, *The Industrial and Commercial Revolution in Great Britain,* 40.

18. Table II.

19. "The increased supplies were chiefly from the West Indies and the Brazils, Turkey giving but little additional assistance. Spinners were quite convinced that unless some new source of supply could be found the progress of the rising industry would be checked if not altogether arrested. Naturally enough it occurred to them that India, the cradle of the cotton industry, would be more likely than any other country to furnish them with their much needed raw material. Accordingly, in 1788, Manchester urged the East India Company to promote the import of cotton from the territories under their jurisdiction. The Company thereupon commenced to bring forward a few parcels by way of experiment, but the quality was so unsuited to the wants of our spinners that it could not be sold except at unremunerative prices" (Ellison, *The Cotton Trade of Great Britain,* 83).

20. Estimate in Stine and Baker, *Cotton Atlas of American Agriculture,* part 5, section A., 8.

21. Buck, op. cit., 34. Quoting from Henry Smithers, *Liverpool,* (Liverpool, 1825), 129, 147.

22. Ellison, op. cit., 89.

23. See Table II.

24. For maps showing the geography of cotton production between 1790 and 1860 see Stine and Baker, op. cit., 16–17, and for the slave distribution during the same period

the maps in F. V. Emerson, "Geographic Influences in American Slavery," *Bulletin of the American Geographical Society* 43 (January–March 1911) between pp. 180–181, will be found interesting.

25. Furnished by Professor Robert E. Park.

26. The reactions against Thomas Dabney, who removed from Virginia to Mississippi in 1835, will illustrate: "It was the custom among the smaller farmers in his neighborhood to call on each other to assist when one of them built his house, usually a log structure. Accordingly, one day an invitation came to the newcomer to help a neighbor 'raise' his house. At the appointed time [Dabney] went over with twenty of his men, and did not leave till the last log was in place and the last board nailed on the room, handing over the simple cabin quite completed to the owner. This action, which seemed so natural to him, was a serious offence to the recipient, and, to his regret, he was sent for to no more 'house-raisings.' On another occasion, a small farmer living a few miles from him got 'in the grass,' as the country people express it when the grass has gotten ahead of the young cotton plants and there is danger of their being choked by it. Again Thomas went over with twenty men, and in a few hours the field was brought to perfect order. The man said that if Colonel Dabney had taken hold of a plough and worked by his side he would have been glad to have his help, but to see him sitting upon his horse with his gloves on directing his negroes how to work was not to his taste. He heard a long time after these occurrences that he could have soothed their wounded pride if he had asked them to come over to help him raise his cabins. But he could not bring himself to call on two or three poor white men to work among his servants when he had no need of help" (Smedes, *Memorials of a Southern Planter,* 29–30).

27. Ingraham, *The Southwest,* 86.

28. Bancroft, op. cit., 351. That western planters had good reason for paying such high prices for slave help is shown by a letter written by a Louisiana planter in 1836: "Many splendid fortunes have been made here in the past three years and many can be made buying these wild lands, clearing them and selling them. Land can be bought for $2.50 an acre, clearing and resold for $20 or $30. This is done daily. My former neighbor, John Weems, went to New Orleans about a month ago to close the purchase of a place for which he was to give $300,000. This sounds big to the ears of a Maryland tobacco planter, but here it is not considered anything. Several places and negroes have been sold since I came here for upwards of $300,000. I was offered one some months past, for which they only asked $500,000, and this was considered cheap at that. Anything under an hundred thousand dollars scarcely takes the attention of a Mississippi River cotton planter. I go on a slower, though perhaps not more sure plan than they do, for where the means is adequate to the purchases, it is quite as easy to pay the one as the other." Quoted in Saxon, op. cit., 137.

29. See Table III.

30. Fleming, op. cit., volume 5, 112.

31. Op. cit., 115.

32. That the free Negroes, at least on the new frontiers, were mainly mulattoes, the result of sex relations between masters and slaves, and originating within the community itself is shown by Sydnor's study of the free colored population in Mississippi. According to this writer, "of the 773 free persons of color in Mississippi in the year 1660, 601 were

of mixed blood, and only 172 were black" (Sydnor, "The Free Negro," 787). Concerning the anomalous position of the free negroes who were usually mulattoes, Phillips says: "Ranging as they did in complexion from a tinged white to full black, in costumes from Parisian finery to many-colored patches, in culture from serene refinement to sloven superstitious uncouthness, these people showed a diverse reflection of the patterns presented by the other groups in the community. Originating nothing, they complied in all things that they might live as a third element in a system planned for two" (op. cit., 172).

33. Table III.
34. Fleming, op. cit., 118.
35. See Barrow, "A Georgia Plantation," 830–836.
36. See Bizzell, *Farm Tenantry in the United States,* and Banks, *Economics of Land Tenure.*
37. Gaines, *The Southern Plantation.*
38. Reuter, *The Mulatto in the United States.*
39. Washington, *Up from Slavery,* 221–222.

6. The Natural History of the Plantation

1. "The sociological point of view makes its appearance in historical investigation as soon as the historian turns from the study of 'periods' to the study of institutions. The history of institutions, that is to say, the church, economic institutions, political institutions, etc., leads inevitably to comparison, classification, the formation of class names or concepts, and eventually to the formulation of law. In the process, history becomes natural history, and natural history passes over into natural science. In short, history becomes sociology" (Park and Burgess, op. cit., 16).
2. Bowman, *The Pioneer Fringe,* 299.
3. Ibid., 299–300.
4. Park, "Our Racial Frontier on the Pacific," 192.
5. Ibid., 196.
6. Gras, *Introduction to Economic History,* passim.
7. "The market areas for most of the products of common use extend far beyond the regions of production. Usually many places compete in supplying the market with a given product. In the competitive situation change in distance, either time or cost, frequently is sufficient to determine the prosperity or decline of a region. That is why there is so much concern about the time schedule and freight races" (McKenzie, "Spatial Distance," 542–543).
8. Ibid., 536.
9. McKenzie, "Ecological Succession in the Puget Sound Region"; Condliffe, *New Zealand in the Making;* and Lind, "Economic Succession and Racial Invasion in Hawaii."
10. Jenks, *The Migration of British Capital to 1875;* Sparks, *History and Theory of Agricultural Credit.*
11. "Migration within the present capitalistic order of machine production and world markets is governed by principles quite different from those which controlled human movement in the past. In the new order of things capital precedes rather than follows settlement, and the regions into which capital now flows are those which contain possibilities for productive capital investment" (McKenzie, *The Evolving World Economy,* 58).

12. John W. Brown makes the following distinction between collective migration and group migration: "*Collective* emigration is the emigration of large numbers of workers generally recruited by agents representing a state, an industry or a concern. Collective emigration is so organized that all the emigrants travel together, and are throughout in the charge of the recruiting organizations or agents appointed by charge of the recruiting organizations or agents appointed by them, by whom they are distributed at their journey's end.

"*Group* migration is a term usually applied to groups or families or co-religionists emigrating for the purpose of land settlement but it is also used to indicate the organization of emigrant groups, so constituted that they will form an independent self-supplying community: for instance, an emigration group of this kind would be provided with a carpenter, a teacher and so on" (*World Immigration and Labour*, 2).

13. After the suppression of the African slave trade a traffic in Chinese and Indian coolies—the so-called "pig business"—began to assume large proportions in the supply of plantation needs. In Cuba, the sugar plantations have imported, besides Chinese coolies, thousands of Negroes from Haiti and Jamaica. There were 63,000 Negroes from these islands in Cuba in 1920 most of whom were recruited by labor agents who are said to have received $15 or $20 each for them from the sugar planters. It is reported that they were herded in plantation barracks with armed guards to prevent them from escaping. At that time their wages averaged from 60 to 80 cents a day (Roller, "El marfil negro," 281–286). With the discontinuation of convict labor in Australia, South Sea Islanders, called Kanakas, proved themselves suited for plantation work. The demand for them, however, was greater than the supply by voluntary enlistment and a new kind of slave trade called "black-birding" arose (Muntz, *Race Contact*, 159–161). In Mexico the "henequen plantations of Yucatan, the coffee plantations of Vera Cruz and Chiapas and the sugar plantations of Morelos used large numbers of workers who came in 'cuadrillas'—gangs from long distances." The Yaqui Indians, in consequence of a rebellion against despoliation of their lands, were transported by the hundreds to Yucatan estates (Tannenbaum, *The Mexican Agrarian Revolution*, 112, 377). The plantations of Hawaii have had a long succession of imported labor groups including Portuguese, Japanese, Chinese, and Filipinos (Lind, op. cit., chapter 4). In Assam, Santhals from other parts of India are recruited to work on the tea plantations. These plantations seem to be unusual in that they employ a large number of "unattached" women, mostly widows or married women living apart from their husbands. With the exception of this unusual sexual ratio, the character of this labor seems to run true to form. "Many of the women have been 'out-caste' for some offense, and quite a considerable proportion are of doubtful reputations. In many cases loose living is deliberately encouraged by the Sirdars or Sirdarins" (Anstey, *The Economic Development of India*, 122).

14. Mackaye, *The New Exploration: A Philosophy of Regional Planning*, 45.

15. Park, op. cit., 193. See also Muntz, op. cit., passim, and Rivers, *Essays on the Depopulation of Melanesia*, passim.

16. An excellent summary of the literature on the question will be found in Ripley, op. cit., chapter 21, "Acclimatization."

17. As Alfred Wallace observed, "the Englishman who can spend a summer in Rome in safety only by sleeping in a Tower and by never venturing forth at night, cannot be truly said to be acclimated" (Quoted in ibid., 586).

18. But Englishmen of the seventeenth century discussed the difficulties of acclimatization in Ireland, see Blount, *Essays on Several Subjects,* 65, and Jacob, *An Historical Account of the Lives,* 87.

19. "Vital conditions do not permit of the accomplishment of plantation labors at the hands of an unacclimated race" (Keller, op. cit., 10).

20. Rivers, op. cit., 89. Essentially the same idea is expressed by Draper: ". . . disease arises from the interplay of dynamic forces inherent in the individual and present in the world about him. It is a subtly moving, changing set of reactions between man and his environment, which cause him discomfort, or exactly what the world says, dis-ease. These discomforts, which have come to be called subjective symptoms, have so often been associated with certain outwardly visible or objective signs, that groupings of these observed phenomena have been described" ("Biological Philosophy and Medicine," volume 1, 123).

21. "It is an interesting fact that of the twenty negroes who were imported in 1619, the first who had arrived in Virginia, not one had died previous to 1624, an indication of the ease with which they stood the deleterious influences of the climate. There was at this time no parallel instance in the history of the white servant" (Bruce, op. cit., 107). See also Wertenbaker, *The Planters of Colonial Virginia,* 128. If white Europeans introduced diseases which depleted the Indian population, Negro slaves introduced diseases which fought on his side in his biological struggle with the whites, and especially the poor whites. Among these were hookworm (the particular foe of poor whites), a mild type of small pox, and possibly malaria and yellow fever (Stitt, "Our Disease Inheritance from Slavery," 801–817). See also Stiles, *Report upon the Prevalence and Geographical Distribution of Hookworm Disease in the United States*; Muntz, op. cit., chapters 10, 19; and 31. Macleod, *The American Indian Frontier,* chapter 4.

22. Herrick, op. cit., 23–24.

23. See an account of the efforts made by the United Fruit Company on their banana plantations in the West Indies and in Central American in Reynolds, *The Banana,* 151.

24. Sombart, *The Quintessence of Capitalism,* 304.

25. Keller, op. cit., 11.

26. Park, "The Mentality of Racial Hybrids," 534.

27. Gillespie, *The Influence of Overseas Expansion on England to 1700,* 47–48.

28. Reynolds, op. cit., 94–96. For the importance of radio communication in the marketing of a perishable fruit like the banana, see Cutter, "The Caribbean Tropics in Commercial Transition," 494–507. Coffee, tea and cacao plantations arose with a large popular market for those beverages as a means of reducing the cost of production lower than that of non-plantation producing areas. The plantation production of tea was experimented with by English interests in the Far East in 1833. Aided by the opening of the Suez Canal and the Russian-Japanese War which cut off the Chinese caravan trade, plantation production prospered and centered in Ceylon and Assam. In 1905 the ancient tea-growing country of China sent a commission to British India to study plantation methods (Mac-Laren, *Rubber, Tea and Cacao,* passim). Production of the coconut and oil palm on a plantation scale has accompanied the substitution of vegetable fats in general for animal fat in human food. This change was stimulated by the food conservation measures of the World War period. In addition, the growing need for cheaper lubricants for machinery and railways contributed to the rise in importance of vegetable oils. The oil palm plantation is the

most recent of the large plantation industries, see von Engeln (op. cit., 286–287). The demand for rubber developed progressively with its use to erase pencil marks, to water-proof cloth, and, after the invention of vulcanizing, for manufacturing pneumatic bicycle tires and finally for automobile tires. The production of rubber took a plantation form when cultivation in the Far East began to be substituted for the exploitation of the indigenous trees in Brazil and in Africa (Akers, op. cit., passim). Hemp and henequen plantations are related to the opening up of vast wheat-producing areas in various parts of the world and the resulting demand for binder twine to be used in harvesting.

29. "A mine owner whose men go out on strike is, briefly, placed in this position: He will lose a sum of money somewhat larger than the amount of profit he could have made during the period of the strike had it not occurred. His coal, however, is still there, and is not less valuable—indeed, in the case of a prolonged strike, may actually be more valu-able—when the strike is over; work can easily be resumed where it was dropped, and during the idle days the ordinary running expenses of the mine cease. The greater part of the loss sustained in the instance I have supposed is not out-of-pocket loss, but merely the failure to realized prospective profits.

"On the other hand, a sugar estate in the tropics spends about eight months out of the twelve in cultivating the crop, and the remaining four in reaping and boiling operations. By the time the crop is ready to reap many thousands of dollars have been expended on it by way of planting, weeding, draining, and the application of nitrogenous manures. If from any cause the labor supply fails when the cutting of the canes is about to commence, every cent expended on the crop is wasted; and if for want of labor the canes which are cut are not transported within a few hours to the mills, they turn sour and cannot be made into sugar. It will thus be seen that in the case of sugar-growing a perfectly reliable labor supply is the first requisite.

"The same might be said of the cultivation of tea, coffee, cocoa, spices, and tropical fruits" (Ireland, "The Labor Problem in the Tropics," 4).

30. Concerning the organization of Hawaiian sugar plantations, Lind says: "Set in obvi-ous and necessary contrast with the common mass have invariably been the aristocracy of labor—sugar boilers, field lunas, surveyors and engineers—the management, and the pro-prietors. Although each of these types represents a certain level in the social and economic scale of the plantation, as compared with unskilled laborers, they continue to form the aristocracy of the community, enjoying a set of privileges which are denied to the great mass of the population. The simple dichotomy of society into the unskilled laborers and the plantation aristocracy is the most obvious basis of plantation stratification and one which outweighs all others" (op. cit., 251–252).

31. This is what John R. Commons calls the "machine theory of labor" (*Industrial Goodwill*, 14–16).

32. The Southern planter still retains a considerable measure of personal authority and customary immunity from the authority of the State. A recent study of lynching says of certain counties in the Black Belt: "A tradition in these counties, respected by sheriff and peace-officers as well as by the public, leaves to the planter and his overseer the settlement of any trouble which arises on the plantation among the Negroes themselves or between them and the overseer or planter. Most crimes in these counties are looked upon as labor

troubles to be settled by those who own and control the plantations. As a corollary, to all practical purposes, the sheriff and other peace-officers are the planter's agents" (*Lynchings and What They Mean*, 30–31).

33. "For a theoretic inquiry into the course of civilized life as it runs in the immediate present, therefore, and as it is running into the proximate future, no single factor in the cultural situation has an importance equal to that of the business man and his work.

"This of course applies with equal force to an inquiry into the economic life of a modern community. In so far as the theorist aims to explain the specifically modern phenomena, his line of approach must be from the business man's standpoint, since it is from that standpoint that the course of these phenomena is directed. A theory of the modern economic situation must be primarily a theory of business traffic, with its motives, aims, methods, and effects" (Veblen, *The Theory of Business Enterprise*, 3–4).

34. Op. cit., 282.

35. It has, apparently, begun over again in Soviet Russia. To bring the land into higher uses farms have been "collectivized," and charges of forced labor have been made; see Chamberlin, *Soviet Russia: A Living Record and a History,* chapter 8, "Karl Marx and the Peasant-Sphinx."

36. Bizzell, *The Green Rising,* chapters 3 and 4.

37. See Manniche, "The Rise of the Danish Peasantry," 35, 130, 218.

38. "[I]t is tolerably clear that the most important races of modern Europe, the Teutons, the Kelts, and the Slavs, have passed within historical times from *extensive* agriculture, in which a patch of soil is exhausted and left to lie waste, to *intensive* agriculture, in which, by an alternative succession of crops and fallow, the same land is used in perpetuity" (Jenks, *Law and Politics in the Middle Ages,* 149).

39. Culbertson, "Raw Materials and Foodstuffs in the Commercial Policies of Nations." See Evans, *The Agrarian Revolution in Roumania,* for an account of the agricultural changes in Roumania since the dissolution of the large estates.

40. See Pratt, *International Trade in Staple Commodities,* chapters 7, 9.

41. In the following passages [about the] story of Huckleberry Finn, Mark Twain has grasped the fact that this process is not due to altruism or even sympathy but simply to mutual accommodation in a common activity:

> It is a far cry from Emerson's ethereal friendship with its fastidious withdrawal from all personal contact to the friendship of Huckleberry Finn and Negro Jim as they lie sprawling on the raft in the middle of the Mississippi. Neither of them would have understood the high moods of the spirit. Neither of them illustrated the dignity of human nature. One was a specimen of the "poor white trash" as it existed on the great river, and the other was a runaway slave. They had not chosen one another; they had literally been "thrown together" as by a careless Fate. They had shared the same crusts, they had smoked together and fished off the same log, and lied and stolen in the common cause of self-preservation. In all this there was nothing consciously ethical or inspiring. When Huckleberry Finn's conscience did assert itself, it was by way of protest against this friendship. His conscience was vague on most points, but

one thing he knew to be wrong. Whatever other form of stealing might be condoned, he was clear in regard to the heinousness of the sin of stealing a slave from his lawful owner. When he slipped off the raft determined to give the information that would send Jim back to slavery, he felt that he was about to do a noble act.

Then he lost his nerve. He refused to obey the inner monitor and sneaked back to his companion. "I got abroad the raft feeling bad and low, because I knowed very well I had done wrong, and I see it wasn't no use for me to learn to do right: a body that don't get started right when he is little ain't got no show when the pinch comes, there ain't nothing to back him up and keep him to his work, and so he gets beat. Then I thought a minute and says to myself, hold on, s'pose you' done right and give Jim up, would you have felt better than you do now? No, says I, I'd feel bad—I'd feel just the same way I feel now. Well, then, says I, what's the use you learning right when it's troublesome to do right and it ain't no trouble to do wrong, and the wages is just the same? I was stuck. I couldn't answer that. So I reckoned I wouldn't bother no more about it, but after this do whichever comes handiest at the time."

Huckleberry Finn was unable to apologize for the impulse upon which he acted. It seemed to him weakness—which he accepted just as he accepted his other manifold weaknesses. He was used to yielding to temptation, and here was another. He was aware that he ought to give up Jim, and he would have done it if he hadn't known him so well, and if Jim had not trusted him. He could not make up his mind to back on his friend." [Samuel McChord Crothers, introduction to *The Book of Friendship* (New York: Macmillan, 1910), vii–viii].

42. Park, review of *Main Currents in the History of American Journalism,* by William Grosvenor Bleyer, 291.

Bibliography

Akers, C. E. *The Rubber Industry in Brazil and the Orient*. London: Methuen and Company, 1914.

Anstey, Vera. *The Economic Development of India*. London: Longmans, Green and Company, 1929.

Bacon, Francis. "On Plantations." In *Works*, vol. 2. Boston: Houghton Mifflin and Company, 1881.

Ballagh, J. C. *A History of Slavery in Virginia*. Baltimore: Johns Hopkins Press, 1902.

———. *White Servitude in the Colony of Virginia*. Baltimore: Johns Hopkins Press, 1895.

Bancroft, Frederick. *Slave-Trading in the Old South*. Baltimore: J. H. Furst Company, 1931.

Banks, E. M. *Economics of Land Tenure in Georgia*. New York: Columbia University Press, 1905.

Barrow, David C., Jr. "A Georgia Plantation." *Scribner's Monthly* 21 (April 1881): 830–836.

Bassett, J. S. "The Relations between the Virginia Planter and the London Merchant." *American Historical Association Report* 1 (1901): 551–575.

———. *A Short History of the United States*. New York: The Macmillan Company, 1918.

Beals, Carleton. "The Black Belt of the Caribbean." *The American Mercury* 24 (October 1931): 129–138.

Becker, Howard. "Ionia and Athens: Studies in Secularization." Ph.D. thesis, University of Chicago, 1930.

Beer, G. L. *The Old Colonial System, 1660–1754*. 2 vols. New York: The Macmillan Company, 1912.

———. *The Origins of the British Colonial System, 1578–1660*. New York: The Macmillan Company, 1908.

Belaunde, V. A. "The Frontier in Hispanic-America." *The Rice Institute Pamphlet* 10 (October 1923): 202–213.

Beverly, Robert. *The History of Virginia*. 2nd ed. London, 1722.

Bizzell, W. B. *Farm Tenantry in the United States*. Texas Agricultural Station Bulletin 278 (April 1921). College Station, Texas.

———. *The Green Rising*. New York: The Macmillan Company, 1926.

Blount, Thomas Pope. *Essays on Several Subjects*. London, 1691.

Bowman, Isaiah. *The Pioneer Fringe*. New York: American Geographical Society, 1931.

Brown, John W. *World Migration and Labour.* Amsterdam: International Federation of Trade Unions, 1926.

Bruce, P. A. *Economic History of Virginia in the Seventeenth Century.* 2 vols. Richmond, 1907.

———. *Social Life of Virginia in the Seventeenth Century.* Richmond, 1907.

Brunhes, Jean, and Camille Vallaux. *Le Géographie de L'Histoire.* Paris: F. Alcan, 1921.

Buck. N. S. *Development and the Organization of Anglo-American Trade, 1800–1850.* New Haven: Yale University Press, 1925.

Buckland, W. W. *The Roman Law of Slavery.* Cambridge: The University Press, 1908.

Buechel, F. A. *The Commerce of Agriculture: A Survey of Agricultural Resources.* New York: John Wiley and Sons, 1926.

Buer, M. C. *Health, Wealth, and Population in the Early Days of the Industrial Revolution.* London: George Routledge and Sons, 1926.

Burgess, E. W. "The Growth of the City." In *The City,* by R. Park, E. W. Burgess, and R. D. MacKenzie. Chicago: University of Chicago Press, 1925.

Butler, Mann. *A History of Kentucky.* Louisville, 1834.

Cambridge History of American Literature. Cambridge, 1917.

Carlyle, Thomas. *Past and Present.* London: George Routledge and Sons, 1907.

Carr-Saunders, A. M. *Population.* London: Oxford University Press, 1925.

Chamberlin, W. H. *Soviet Russia: A Living Record and a History.* Boston: Little, Brown and Company, 1931.

Channing, Edward. "The Narragansett Planters." *Johns Hopkins University Studies in Historical and Political Science* 4 (1886): 105–127.

Clements, F. E. *Plant Succession and Indicators.* New York: H. W. Wilson Company, 1928.

Coman, Katherine. *The Industrial History of the United States.* New York: The Macmillan Company, 1907.

Commons, John. *Industrial Goodwill.* New York: The McGraw-Hill Company, 1919.

Condliffe, J. B. *New Zealand in the Making: A Survey of Economic and Social Development.* Chicago: University of Chicago Press, 1930.

Conway, M. D., ed. *George Washington and Mount Vernon.* Long Island Historical Society Memoirs 4. Brooklyn: Long Island Historical Society, 1889.

Cooley, C. H. *The Theory of Transportation.* Publications of the American Economic Association 9. [Baltimore]: American Economics Association, 1894.

Craven. A. O. *Soil Exhaustion as a Factor in the Agricultural History of Virginia and Maryland, 1606–1680.* University of Illinois Studies in the Social Sciences 13. Urbana: University of Illinois Press, 1926.

Culbertson, William S. "Raw Materials and Foodstuffs in the Commercial Policies of Nations." *Annals of the American Academy of Political and Social Science* 112 (March 1924): 1–147.

Cutter, Victor M. "The Caribbean Tropics in Commercial Transition." *Economic Geography* 2 (October 1926): 494–507.

Davis, John P. *Corporations: A Study of the Origin and Development of Great Business Corporations and Their Relations to the Authority of the State.* 2 vols. New York: G. P. Putnam's Sons, 1905.

Dawson, C. A., and W. E. Gettys. *An Introduction to Sociology.* New York: The Ronald Press, 1929.

Day, Clive. *A History of Commerce.* New York: Longmans, Green and Company, 1919.

Destler, C. M. "The Tobacco Industry in Virginia, 1783–1860." M.A. thesis, University of Chicago, 1928.

Dew, T. R. *Review of the Debate in the Virginia Legislature of 1831 and 1832.* Richmond, 1832.

Dodd, W. E. *The Cotton Kingdom: A Chronicle of the Old South.* New Haven: Yale University Press, 1921.

———. "The Plantation and Farm Systems in Southern Agriculture." In *The South in the Building of the Nation,* 5:73–80. Richmond: Southern Historical Publication Society, 1909.

Donaldson, John. *International Economic Relations: A Treatise on World Economy and World Politics.* New York: Longmans, Green and Company, 1928.

Douglas, Paul H. "American Apprenticeship and Industrial Education." *Columbia University Studies* 95 (1921): 207–554.

Doyle, J. A. *English Colonies in America: Virginia, Maryland, and the Carolinas.* New York: Henry Holt and Company, 1889.

Draper, George. "Biological Philosophy and Medicine." *Human Biology* 1 (January 1929): 117–135.

Dryer, Charles Redway. "Mackinder's 'World Island' and Its American 'Satellite.'" *Geographical Review* 9 (March 1920): 205–207.

Ellison, Thomas. *The Cotton Trade of Great Britain.* London: Effingham Wilson, 1886.

Emerson, F. V. "Geographic Influences in American Slavery." *American Geographical Bulletin* 43 (January–March 1911): 13–26, 106–118, and 170–181.

Evans, I. L. *The Agrarian Revolution in Roumania.* Cambridge: The University Press, 1924.

Faris, Ellsworth, "The Origin of Punishment." *International Journal of Ethics* 25 (October 1914): 54–67.

Fiske, John. *Old Virginia and Her Neighbours.* Boston: Houghton Mifflin, 1897.

Fleming, Walter Lynwood. "The Slave-Labor System in the Ante-Bellum South." In *The South in the Building of the Nation,* 5:104–120. Richmond: Southern Historical Publication Society, 1909.

Ford, C. W., ed. *The Writings of George Washington.* New York: G. P. Putnam's Sons, 1889–1893.

Fuchs, Carl J. "The Epochs of German Agrarian History and Agrarian Policy." In *Selected Readings in Rural Economics,* edited by T. N. Carver, 223–253. Boston: Ginn and Company, 1916.

Gaines, F. P. *The Southern Plantation: A Study in the Development and Accuracy of a Tradition.* New York: Columbia University Press, 1924.

Gee, Wilson, and John J. Corson. *Rural Depopulation in Certain Tidewater and Piedmont Areas of Virginia.* University of Virginia, the Institute for Research in the Social Sciences, no. 3. University, Va.: Institute for Research in the Social Sciences, 1929.

Gillespie, J. E. *The Influence of Overseas Expansion on England to 1700.* New York: Columbia University Press, 1920.

Gras, N. S. B. *A History of Agriculture in Europe and America.* New York: F. S. Crofts and Company, 1925.

———. *Introduction to Economic History.* New York: Harper and Brothers, 1922.

Gray, L. C. "Southern Agriculture, Plantation System, and the Negro Problem." *Annals of the American Academy of Political and Social Science* 40 (March 1912): 90–99.

Hakluyt, Richard. *The Principal Navigations, Voyages, Traffiques, and Discoveries of the English Nation,* edited by E. Goldsmid. Edinburgh: E. & G. Goldsmid, 1885–1890.

Hammond, J. L., and Barbara Hammond. *The Rise of Modern Industry.* New York: Harcourt, Brace and Company, 1926.

Hening, W. W. *Statutes at Large in Virginia.* Richmond: Printed by and for Samuel Pleasants, 1823.

Herrick, C. A. *White Servitude in Pennsylvania.* Philadelphia: J. J. McVey, 1926.

Ingle, Edward. "Local Institutions of Virginia." *Johns Hopkins University Studies in Historical and Political Science,* 3rd series, 2–3 (1885): 1–127.

Ingraham, J. H. *The Southwest.* New York: Harper and Brothers, 1835.

Ingram, John K. *A History of Slavery and Serfdom.* London, 1895.

Ireland, Alleyne. "The Labor Problem in the Tropics." *Appleton's Popular Science Monthly,* February 1899, 481–490.

———. *Tropical Colonization.* New York: The Macmillan Company, 1899.

Jacob, Giles. A*n Historical Account of the Lives and Writings of Our Most Considered English Poets, Etc.* London, 1724.

Jefferson, Thomas. *Notes on the State of Virginia.* London, 1787.

Jenks, Edward. *Law and Politics in the Middle Ages.* New York: Henry Holt and Company, 1898.

Jenks, Leland H. *The Migration of British Capital to 1875.* New York: A. A. Knopf, 1927.

Jernegan, M. W. "A Forgotten Slavery of Colonial Days." *Harper's Monthly* 77 (October 1912): 745–751.

Johnson, Alvin. "The War—by an Economist." *Unpopular Review* 2 (1914): 411–428.

Johnston, Sir H. H. *The Backward Peoples and Our Relations with Them.* London: H. Milford Oxford University Press, 1920.

Keim, Clarence Ray. "Influence of Primogeniture and Entail in the Development of Virginia." Ph.D. thesis, University of Chicago, 1926.

Keller, A. G. *Colonization: A Study in the Founding of New Societies.* Boston: Ginn and Company, 1908.

Kingsbury, Susan M. "A Comparison of the Virginia Company with the Other English Trading Companies of the Sixteenth and Seventeenth Centuries." *American Historical Association Report* 1 (1906) 161–176.

Knight, Melvin M., Harry E. Barnes, and Felix Flugel. *Economic History of Europe.* Boston: Houghton Mifflin, 1928.

Knowles, L. C. A. *The Economic Development of the British Overseas Empire.* London: G. Routledge and Sons, 1928.

————. *The Industrial and Commercial Revolutions in Great Britain during the Nineteenth Century.* London: G. Routledge and Sons, 1926.

Lauber, A. W. *Indian Slavery in Colonial Times within the Present Limits of the United States.* New York: Columbia University Press, 1913.

Lawson, John. *History of Carolina.* London, 1714.

Leake, H. M. *Land Tenure and Agricultural Production in the Tropics.* Cambridge: W. Heffer and Sons, 1927.

Leroy-Beaulieu, Paul. *De la Colonisation chez les Peuples Modernes.* 5th ed. 2 vols. Paris, 1902.

Lind, Andrew W. "Economic Succession and Racial Invasion in Hawaii." Ph.D. thesis, University of Chicago, 1931.

Lucas, Sir. C. P. *The Beginnings of English Overseas Enterprise: A Prelude to Empire.* Oxford: Clarendon Press, 1917.

Lynchings and What They Mean: General Findings of the Southern Commission on the Study of Lynching. Atlanta: The Commission, 1931.

MacInnes, C. M. *The Early English Tobacco Trade.* London: K. Paul, Trench, Trubner and Company, 1926.

Mackaye, Benton. *The New Exploration: A Philosophy of Regional Planning.* New York: Harcourt, Brace, 1928.

Mackinder, H. J. *Democratic Ideals and Reality: A Study in the Politics of Reconstruction.* London: Constable and Company, 1919.

MacLaren, W. A. *Rubber, Tea and Cacao.* London, 1924.

Macleod, W. C. *The American Indian Frontier.* New York: A. A. Knopf, 1928.

————. "Big Business and the North American Indian." *American Journal of Sociology* 34 (November 1928): 480–491.

————. "Debtor and Chattel Slavery in Aboriginal North America." *American Anthropologist,* n.s., 27 (July 1925): 370–380.

————. "Economic Aspects of Indigenous American Slavery." *American Anthropologist,* n.s., 30 (October 1928): 632–650.

Macpherson, David. *Annals of Commerce.* London, 1805.

Maine, Sir Henry Summer. *Ancient Law.* 14th ed. London: John Murray, 1891.

Manniche, P. "The Rise of the Danish Peasantry." *Sociological Review* 19 (1927): 35–37, 218.

Mantoux, Paul. *The Industrial Revolution the Eighteenth Century.* Rev. ed. Translated by Marjorie Vernon. New York: Harcourt, Brace and Company, 1928.

Marshall, Alfred. *Principles of Economics.* 6th ed. London: Macmillan and Company, 1910.

Maury, Ann. *Memoirs of a Huguenot Family.* New York: G. P. Putnam's Sons, 1872.

McKenzie, R. D. "Ecological Succession in the Puget Sound Region." *Publications of the American Sociological Society* 23 (1929): 60–80.

————. *The Evolving World Economy.* Albert Kahn Foundation for the Foreign Travel of American Teachers 5. New York, 1926.

————. "Food Supply in Relation to Population." *Proceedings of the Institute of International Relations,* 1926.

————. "Spatial Distance." *Sociology and Social Research* 13 (July–August 1929): 536–544.

McLaren, Jack. *My Crowded Solitude.* New York; Robert M. McBride and Company, 1926.

Mead, G. H. "The Philosophies of Royce, James and Dewey in Their American Setting." *International Journal of Ethics* 40 (January 1930): 211–231.

Merivale, Herman. *Lectures on Colonization and Colonies.* London: Longman, Orme, Brown, Green, and Longmans, 1841.

Mooney, James. *The Aboriginal Population of America North of Mexico.* Smithsonian Publication No. 2955. Washington D.C.: The Smithsonian Institution, 1928.

Moret, A., and G. Davy. *From Tribe to Empire: Social Organization among Primitives and in the Ancient East.* New York: A. A. Knopf, 1926.

Munford, Beverley B. *Virginia's Attitude toward Slavery and Secession.* Richmond: New Edition, 1910.

Muntz, E. E. *Race Contact.* New York: The Century Company, 1927.

Murray, Gilbert. *The Rise of the Greek Epic.* Oxford: Clarendon Press, 1907.

Nieboer, H. J. *Slavery as an Industrial System: Ethnological Researches.* 1900. 2nd rev. ed. The Hague: Martinus Nijhoff, 1910.

Oppenheimer, Franz. *The State: Its History and Development Viewed Sociologically.* Translated by J. M. Gitterman. Indianapolis: The Bobbs-Merrill Company, 1914.

Park, Robert E. "Education and Its Relation to the Conflict and Fusion of Cultures." *Publications of the American Sociological Society* 13 (December 1918): 38–63.

————. "Human Migration and the Marginal Man." In E. W. Burgess, ed., *Personality and the Social Group.* Chicago: University of Chicago Press, 1929.

————. *The Immigrant Press and Its Control.* New York: Harper and Brothers, 1922.

————. "The Mentality of Racial Hybrids." *American Journal of Sociology* 36 (January 1931): 534–551.

————. "Our Racial Frontier on the Pacific." *The Survey* 56 (May 1, 1926): 192–194.

————. Review of *Main Currents in the History of American Journalism,* by William Grosvenor Bleyer. American Journal of Sociology 33 (September 1927): 291.

————. "The Urban Community as a Spatial Pattern and a Moral Order." In E. W. Burgess, ed., *The Urban Community.* Chicago: University of Chicago Press, 1926.

Park, Robert, and Ernest W. Burgess. *Introduction to the Science of Sociology.* 2nd ed. Chicago: University of Chicago Press, 1924.

Parks, George Bruner. *Richard Hakluyt and the English Voyages.* New York: American Geographical Society, 1928.

Penn, William. *Some Account of the Province of Pennsylvania.* Pamphlet originally published in 1681.

Perry, W. J. *The Children of the Sun: A Study in the Early History of Civilization.* London: Methuen, 1927.

Phillips, U. B. *American Negro Slavery.* New York: Appleton Company, 1918.

————. "The Decadence of the Plantation System." *Annals of the American Academy of Political and Social Science* 35 (January 1910): 37–41.

———. "The Economics of the Slave Trade: Foreign and Domestic." In *The South and the Building of the Nation,* 5:124–129. Richmond: Southern Historical Publication Society, 1909.

———. *Life and Labor in the Old South.* Boston: Little, Brown and Company, 1929.

———. *Plantation and Frontier Documents, 1649–1863.* Vols. 1 and 2 of *A Documentary History of American Industrial Society,* edited by J. R. Commons et al. Cleveland: The Arthur H. Clark Company, 1909–1910.

Pratt. E. E. *International Trade in Staple Commodities.* New York: McGraw-Hill Book Company, 1926.

Raleigh, Walter. *Romance: Two Lectures.* Princeton: Princeton University Press, 1916.

Reinsch, Paul S. *Colonial Government: An Introduction to the Study of Colonial Institutions.* New York: The Macmillan Company, 1906.

Reuter, E. B. *The Mulatto in the United States: Including a Study of the Role of the Mixed-Blood throughout the World.* Boston: Richard G. Badger, 1918.

Reynolds, Philip K. *The Banana: Its History, Cultivation and Place among Staple Foods.* Boston: Houghton Mifflin, 1927.

Ripley, W. Z. *The Races of Europe.* New York: D. Appleton and Company, 1919.

Rivers, W. H. R. *Essays on the Depopulation of Melanesia.* Cambridge: The University Press, 1922.

Roller, Arnold. "El marfil negro y el oro blanco en Cuba." *Riviste Bimestre Cubana* 25 (March–April 1930); 281–286.

Roscher, Wilhelm, and Robert Jannasch. *Kolonien, Kolonialpolitik, und Auswanderung.* Leipzig: C. F. Winter, 1885.

Rose, J. H., A. P. Newton, and E. A. Benians. *The Cambridge History of the British Empire.* New York: The Macmillan Company, 1929.

Russell, J. H. *The Free Negro in Virginia, 1619–1865.* Baltimore: Johns Hopkins Press, 1913.

Saxon, Lyle. *Old Louisiana.* New York: The Century Company, 1929.

Schmidt, Louis B., and E. D. Ross. *Readings in the Economic History of American Agriculture.* New York: The Macmillan Company, 1925.

Scisco, L. D. "The Plantation Type of Colony." *The American Historical Review* 8 (January 1903): 260.

Shepherd, William R. "The Expansion of Europe." Pts. 1, 2, and 3. *Political Science Quarterly* 34 (March 1919): 43–60; 34 (June 1919): 210–225; 34 (September 1919): 392–412.

Slavery and the Internal Slave Trade in the United States, Being Replies to Questions of the Committee of British and Foreign Anti-Slavery Societies. London: Executive Council of the American Anti-Slavery Society, 1841.

Smedes, Susan Dabney. *Memorials of a Southern Planter.* Baltimore: Cushings and Bailey, 1887.

Smith, J. Russell. *Industrial and Commercial Geography.* New York: Henry Holt and Company, 1913.

———. *North America: The People and the Resources.* New York: Harcourt, Brace, and Company, 1925.

Smith, John. *Works: 1608–1631*. [Birmingham: Edward Arber,] 1884.

Sombart, Werner. *The Quintessence of Capitalism*. New York: E. P. Dutton & Company, 1925.

Sorokin, P. A., C. C. Zimmerman, C. J. Galpin, eds. *A Systematic Source Book in Rural Sociology*. Vol. 1. Minneapolis: University of Minnesota Press, 1930.

Spann, Othmar. *The History of Economics*. Translated from the 19th German edition by Eden and Cedar Paul. New York: W. W. Norton and Company, 1930.

Sparks, Earl Sylvester. *History and Theory of Agricultural Credit in the United States*. New York: Thomas Y. Crowell Company, 1932.

Spengler, Oswald. *The Decline of the West*. 2 vols. Authorized translation with notes by C. F. Atkinson. New York: A. A. Knopf, 1928.

Spinder. H. J. "The Population of Ancient America." *Geographical Review* 18 (1928): 641–660.

Stiles, C. W. *Report upon the Prevalence and Geographic Distribution of Hookworm Disease in the United States*. Treasury Dept., Hygienic Laboratory Bulletin No. 10. Washington, 1903.

Stine, O. C., and O. E. Baker. *Cotton Atlas of American Agriculture*. Part 5, section A. Washington: Government Printing Office, 1918.

Stitt, E. R. "Our Disease Inheritance from Slavery." *United States Naval Medical Bulletin* 26 (October 1928): 801–817.

Sumner, W. G. "Advancing Social and Political Organization in the United States." In *The Challenge of Facts and Other Essays*. New Haven: Yale University Press, 1914.

———. *Folkways*. Boston: Ginn and Company, 1906.

Sydnor, Charles S. "The Free Negro in Mississippi before the Civil War." *American Historical Review* 32 (July 1927): 771–779.

Tannenbaum, Frank. *The Mexican Agrarian Revolution*. New York: The Macmillan Company, 1929.

Teggart, F. J. *The Processes of History*. New Haven: Yale University Press, 1918.

———. *Theory of History*. New Haven: Yale University Press, 1925.

Thomas, W. I., and Florian Znaniecki. *The Polish Peasant in Europe and America*. 5 vols. Boston: Richard G. Badger, 1920.

Thünen, J. H. von. *Der isolierte Staat in Beziehung auf Landwirtschaft und National-ökonomie*. 2nd ed. Rostock, 1842–1865.

Traquair, Ramsay. "The Commonwealth of the Atlantic." *The Atlantic Monthly* 133 (May 1924): 602–608.

Trotter, W. "Herd Instinct—Part 1." *Sociological Review* 1 (July 1908): 227–248.

Tschan, F. J. "The Virginia Planter, 1700–1775." Ph.D. thesis, University of Chicago, 1916.

Turner, F. J. *The Frontier in American History*. New York: Henry Holt and Company, 1920.

University of Virginia News Letter. 6 (February 1, 1930).

Veblen, Thorstein. *The Theory of Business Enterprise.* New York: Charles Scribner's Sons, 1904.

von Engeln, O. D. *Inheriting the Earth.* New York: The Macmillan Company, 1922.

Washington, Booker T. *Up from Slavery: An Autobiography.* New York: Doubleday, Page and Company, 1903.

Wertenbaker, R. J. *The First Americans, 1607–1690.* New York: The Macmillan Company, 1927.

———. *Patrician and Plebeian in Virginia, or The Origin and Development of the Social Classes of the Old Dominion.* Charlottesville: University of Virginia Press, 1910.

———. *The Planters of Colonial Virginia.* Princeton: Princeton University Press, 1922.

Westermarck, Edward. *The Origin and Development of the Moral Ideas.* London: Macmillan and Company, 1912–1917.

Woofter, T. H., Jr. *Black Yeomanry: Life on St. Helena Island.* New York: Henry Holt and Company, 1930.

World Agricultural Census of 1930. Rome: International Institute of Agriculture, 1930.

Index

aboriginal Indians, 37, 39, 40–41, 43, 44–46; European perceptions of, 40, 43, 49; and plantation labor, 22, 121n18, 122n23; and use of slaves, 41

acculturation, 37

"adventurers of the person," 71

African Americans. *See* "Negroes"

African Company, 42

agrarian revolution, 23–24, 83, 85

agricultural reform, 85

American exceptionalism, xii

American South, ix, x, xii, 22; the cultural order of, 99, 100; and Europe, xii, 74, 100; romantic images of, 98

anthropology, xiii–xiv

apprenticeship, 57–58, 76

arboriculture, 20, 86

Atlantic trade, 23–24

Bacon, Francis: *On Plantations,* 20–21

Beverly, Robert, 72

black leaders, 99

black slave owners, 67

Boas, Franz, xii, xiii

Brazilian frontier, 101–2

Brown, William Oscar, xi

Burke, Edmund, 38

California frontier, 71

Casor, John, 67

Christianization of slaves, 71

civilization, xi, 9, 11, 15, 19, 20, 38, 86, 113n5, 117n56, 119n16

climatic theories of the plantation, 5–6, 105

coastlands, 27

coerced labor, 7, 8, 10, 13

colonialism, xii

colonization, xii, 4–8, 42, 43; and the world market, 3, 4

commercial factory in Virginia, 32–33

commercial revolution, 24

conquest, x, 2, 9, 14, 15, 16, 23, 43, 46

corporal punishment, 75

cotton, 87–88, 89; and conquest 88–89; expansion of cotton plantations, 88–89

cotton frontier, 93–94; and the distribution of slaves, 94

cotton gin, 87, 88

cropper tenantry, 98

demagogues, 98–99

Dillon, S.C., x

disease, 43, 105, 133n20, 133n21

distance, revolutions of, 23–26

division of labor, 14, 19, 37, 38, 50, 73–74, 103, 108, 120n38; and free "Negroes," 80; and race, 103, 105

domestic slave trade, 89, 94–95

Duke University, xii

Dutch West India Company, 35

Eastern Prussia, 7

East India Company, 31, 41–42, 43

Eastland Company, 30–31

Elizabeth I (queen), 39–40

emancipation, 17–18

emigration, 84, 85, 132n12
Enclosure Movement, 24
English maritime enterprises, 33
English merchant class, 72
English Navigation Acts, 54
English Poor Laws, 25
European state power, x

Fields, Barbara J., xvi, xvii
financial revolution, 23–24, 118n4
Fitzhugh, William, 47, 72
Frazier, E. Franklin, xi
free "Negroes," 67, 79–80, 95; as a threat,
    80
frontier, x, xv, xvi, xvii, 1, 2, 7, 8, 9–10,
    18, 89, 94, 101–3, 104, 107, 109, 110,
    111; conditions of, 65; and expansion
    of, 87–99; and the plantation, 18, 64,
    82–87

German colonization, 7
globalization, ix, xvi. See also world
    community
Greek polis, xvi, 9–10

Hakluyt, Richard, 33
Hanseatic League, 30
headright, 47–48, 121n20
Henry IV (king), 30
Hindu coolies, 18
history, study of, 2
homicide of slaves, 75, 76
house-father, Roman, 75–76
Hughes, Evertt C., xi
Hughes, Helen MacGill, xi
hybridization, 106

immigration, 56, 72.
indentured servitude, 47, 57–60, 65, 76;
    and control of industrial labor, 60;
    involuntary, 58–59; punishment and,
    61–63. See also headright

Indianization, 37
Indian massacre (of 1622), 49, 60
Indian War of 1676, 43
individualism, 10, 38, 106
individualization, 9, 19, 21; and proletari-
    anization, 9
industrial revolution, 24
infanticide, 16
intensive agriculture, 19, 86, 87, 110,
    135n38
interracial sex, 69
interracial social relations, xii, 20, 111

Jamestown, 44–45
Jim Crow South, xi–xii
Johnson, Anthony, 66–67
Johnson, Charles S, xi
joint-stock companies, 29, 31, 41
joint-stock plantations, 47

Keller, A. G., 5–6, 7, 105, 106, 114n23;
    and climatic theory of the plantation,
    7–8, 104–5, 111n23; and control of
    labor, 108–9

labor shortages, 68, 76
labor supply, 11–13, 104, 106

Mackinder, H. J., 27
Maine, Henry, xii, 2
Malthus, Thomas, 16
manor, 3–4
manumitting slaves, 78
Marx, Karl, xii
merchant-adventurers, 30–31, 119n12
merchant-planters, 72, 73, 127n49
merchants, 28, 29, 30, 31, 34, 41, 53,
    55, 72, 83, 126n28; of Germany, 7;
    of England, 30, 31, 34, 41, 53; of
    Virginia, 42, 44
migration, 26, 35, 57, 104, 106. See also
    immigration

miscegenation, 69, 106, 11

Montgomery, Isaiah, 89–90

More, Thomas: *Utopia,* 26

Morgan, Lewis Henry, xii, xv

Mound Bayou, Miss., 89

Murray, Gilbert, 9–10

Muscovy Company, 31, 41

nationalism, 14

Native Americans, 37, 40–41, 43, 44–46

"Negroes": in competition with white labor, 64; free, 67, 79–80, 95; growth in American population of, 70; as indentured servants, 64–66; and slavery, 17–18, 63–71, 81, 94

New World, 23; and precious metals, 23

Nieboer, Herman Jeremias, xv, 11–12, 16, 23, 118n5; and open resources, 12–13, 39

ocean transport, 5

On Plantations (Bacon), 20–21

open resources, xvii, 3, 12, 15, 16, 19, 21, 22, 39, 103, 104

Oppenheimer, Franz, xv, 14–15, 16, 28, 110

Park, Robert Ezra, xi, 102–3, 104, 106, 111, 113n5, 114n23

partition of Poland, 14

peasant proprietorship, 8, 19, 95, 98, 103, 110; and cultural homogeneity, 110–11

Penn, William, 25

plantation aristocracy, 71–72

plantations, x, xiii, 3–4, 20–21; climatic theories of, 2, 5, 6, 105; and colonization, xii, 4–8; and conquest, x, 2, 15, 43, 87; decline of, 74, 79–80, 82–83, 86, 95–98; distribution of, 5–6; and division of labor, 37–38, 104; and frontier, 18, 89, 95, 101–2; as a frontier institution, xv, 1, 2, 7, 8, 10, 18, 89,

109; government of, 36–37, 74–75; humanization of, 74–78; as ideal society, 26; as industrial system, 5, 8, 51, 56, 60, 66, 77, 111, 134n30; as institution for settlement, 87, 109; and migration, 26; natural history of, xiii, 110; as a political institution, x, xv, xvi, 1–3, 10, 13–15, 109; and slavery, xii, xiii, 23, 72; and social change, 18–20; state making, xv, 1; theory of, 1, 15–18

plantations in temperate regions, 7–8

plantation slavery as industrial system, 5, 8, 51, 56, 60, 66, 77, 111

planter-adventurers, 72

planters, 2, 18, 22, 33–34, 41, 42, 44, 48–49, 53–54, 126n42; authoritarian power of, xv, 2, 75–77, 109; definition of, 71; and English society 74; evolution of, 71–74; and government, 76; and manumission, 78–79 ; personality of, 73–74

population changes, 96–97

population growth, 70

population pressure, 16

power farming, 4, 77

Powhatan, 40–41, 44–45

private gardens, 46

provisioning of food, 82

race, xi, xiii, xvi–xvii, 20, 64, 70–71; and division of labor, 14, 103, 105, 108, 109; imputing characteristics of, xiii, 105; and labor, 64, 65; and labor control, xiii, xvii; social construction of, x, xvi–xvii; and stratification, 108

race relations, 71, 102, 103, 104–5, 111

racial determinism, 2, 3

racial stratification, 107–8

Redfield, Robert, xi, xii, xv

Reinsch, Paul, 11

religion, 66

Roman house-father, 75–76

romantic images of the antebellum South, 98

Royal African Company, 72

Royce, Josiah, 71

Ruffin, Edmund, 85

scientific objectivity, xi

sea nomads, 28

sharecropping, 98

slavery, 11–13, 63–71, 76, 89, 94, 104, 124n21; humanization of, 78–81; as an industrial system, 1; justified by race, 70, 71; and labor shortages, 13, 173; as a society of open resources, 12, 104; transition to, 64–67

slaves, 65; homicide of, 75, 76; and inheritance, 78–79

slave trade, 76, 89, 94

social change, 2, 18–20; and evolving social institutions, 18–19; the unimportance of biology in, 18

soil exhaustion, 83–84, 85

Sombart, Werner, xii

South America, 11

southern exceptionalism, x

spatial relations, 26–27

state, the, x; as institution of settlement, 14; origins of, 13–14; and slavery 14

state making (state-making), xv, 1, 15

Steward, Julian, xiii–xiv

Sumner, W. G., 37

Tacitus, 39

Teggart, Frederick, 1–3, 14, 19

textile innovations, 88, 90–93

tobacco, 46, 49–55, 82, 83; marketing of, 53–55, overproduction of, 53–54; and rationalization of labor, 76–77

trading company, 24–25, 29, 31

trading factory, 20, 33–38, 44–45, 54–55; in the Old World, 44; transition to plantations, 33–38

Turner, Frederick Jackson, 10, 14

Tuskegee Institute, 99

uneven development, 2, 103

Utopia (More), 26

Venice Company, 31

Virginia, 39–55; as typical plantation frontier, 20–22

Virginia Company, 34, 41–44, 57, 71

Virginia Slave Code (1662), 68

Washington, Booker, T., xi, 99

Washington, George, 85

Weber, Max, xii

white labor, 56–63

white supremacy, xii

Whitney, Eli, 88

wool trade, 28–29

world community, x, xv, 3, 4, 5, 7, 8, 56, 103, 107, 109, 131n11; England as center of, 21

World Island, 27, 32

world markets, 3, 4, 5, 11, 107, 109, 131n11

World Ocean, 27

World War II, xiii